Metal Organic Frameworks: Synthesis and Application II

Metal Organic Frameworks: Synthesis and Application II

Guest Editors

Victoria Samanidou
Eleni Deliyanni

MDPI

Basel • Beijing • Wuhan • Barcelona • Belgrade • Novi Sad • Cluj • Manchester

Guest Editors
Victoria Samanidou
School of Chemistry
Aristotle University of Thessaloniki
Thessaloniki
Greece

Eleni Deliyanni
School of Chemistry
Aristotle University of Thessaloniki
Thessaloniki
Greece

Editorial Office
MDPI AG
Grosspeteranlage 5
4052 Basel, Switzerland

This is a reprint of the Special Issue, published open access by the journal *Molecules* (ISSN 1420-3049), freely accessible at: https://www.mdpi.com/journal/molecules/special_issues/MOF_Syn_app.

For citation purposes, cite each article independently as indicated on the article page online and as indicated below:

Lastname, A.A.; Lastname, B.B. Article Title. *Journal Name* **Year**, *Volume Number*, Page Range.

ISBN 978-3-7258-6894-0 (Hbk)
ISBN 978-3-7258-6895-7 (PDF)
https://doi.org/10.3390/books978-3-7258-6895-7

Contents

molecules

MDPI

Article

Adsorptive Elimination of a Cationic Dye and a Hg (II)-Containing Antiseptic from Simulated Wastewater Using a Metal Organic Framework

Nilanjan Roy [1], Chanchal Das [1], Mohuya Paul [2], Jungkyun Im [2,3,*] and Goutam Biswas [1,*]

1 Department of Chemistry, Cooch Behar Panchanan Barma University, Cooch Behar 736101, West Bengal, India; roynilanjancob@gmail.com (N.R.); chanchaldas453@gmail.com (C.D.)
2 Department of Electronic Materials, Devices and Equipment Engineering, Soonchunhyang University, Asan 31538, Republic of Korea; mohuyapaul678@gmail.com
3 Department of Chemical Engineering, Soonchunhyang University, Asan 31538, Republic of Korea
* Correspondence: jkim5279@sch.ac.kr (J.I.); goutam@cbpbu.ac.in (G.B.)

Abstract: Several types of pollutants have acute adverse effects on living bodies, and the effective removal of these pollutants remains a challenge. Safranin O (a biological dye) and merbromin (a topical mercury-containing antiseptic) are considered organic pollutants, and there are only a few reports on their removal. Synthesized and well-characterized (through PXRD, FTIR, FESEM, and EDS analysis) MOF-5 was used for the first time in the removal of safranin O and merbromin from simulated wastewater and real wastewater. In both cases, MOF-5 effectively removed contaminants. We found that in simulated wastewater, the highest efficiency of removal of safranin O was 53.27% (for 15 mg/L) at pH 10, and for merbromin, it was 41.49% (for 25 mg/L) at pH 6. In the case of real wastewater containing natural ions (Na^+, K^+, F^-, Cl^-, SO_4^{2-}, PO_4^{3-}, Mg^{2+}, and Ca^{2+}) and other molecules, the removal efficiencies of these two dyes decreased (34.00% and 26.28% for safranin O and merbromin, respectively) because of the presence of other ions and molecules. A plausible mechanism for the removal of these pollutants using MOF-5 was proposed.

Keywords: metal organic framework (MOF); wastewater management; natural ions; safranin O and merbromin; dyes and pharmaceutical wastes; adsorption

1. Introduction

The amount of pure drinkable water decreases daily, for both natural and artificial reasons. Industrial development is a major contributor to water pollution. Since the very beginning of the industrial revolution, as the industrial sectors have expanded every year, the amount of various hazardous chemicals contaminating groundwater has increased rapidly [1,2]. Different types of dyes are pollutants that add up to a serious issue of water pollution [3]. There are various types of dyes that are used in different industries, especially in the textile industry, broadly divided according to the (a) method of application and (b) chemical structure [4]. On the other hand, based on charge, dyes can be classified into three classes: (a) cationic dyes (e.g., rhodamine B, methylene blue, and rhodamine 6G), (b) anionic dyes (e.g., Congo red, methyl orange, and acid fuchsin), and (c) neutral dyes (e.g., Giemsa stain) [5,6]. In general, anionic dyes tend to be acidic, whereas cationic dyes are basic in nature [7]. Anionic dyes are utilized for the modification of acrylic, polyamide, and polypropylene fibers, whereas cationic dyes are frequently used in the dyeing of acrylic, wool, and silk fibers [5]. There are also different pharmaceutical pollutants, such as staining dyes, antibiotics, antimicrobial medicines, and other harmful drugs, which are also responsible for serious water pollution and severe negative impacts on the environment. Safranin-O and merbromin (Figure 1) are hazardous pollutants [8]. Safranin O (or safranin) is a well-known cationic dye that is frequently used to stain cells in laboratories and hospitals [9]. However, safranin O also has detrimental effects on the cardiovascular system and other serious health problems, including methemoglobinemia, cyanosis, spasms, and severe eye and skin irritation [10].

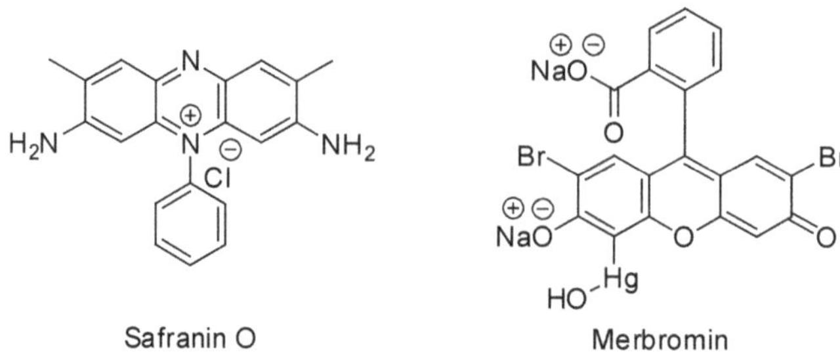

Figure 1. Structures of safranin O and merbromin.

On the other hand, merbromin, a mercury-ion containing pharmaceutical waste (which is a sodium salt of 2,7-dibromo-4-hydroxymercurifluorescein), is typically used as an antiseptic, mostly for external treatments of cuts and scrapes [11]. Although many of us are unaware of its hazardous effects, we apply it to different injuries, minor wounds, and burns [11]. It is responsible for gastrointestinal problems, such as nausea and vomiting like some other drugs and pharmaceuticals [12]. Merbromin can result in high mercury concentrations in the blood, urine, and other organs, as well as severe digestive system damage, renal failure, and toxicity to other organs [11].

Therefore, it is necessary to eliminate safranin O and merbromin from water to prevent their adverse effects on many living systems. Safranin O and merbromin were chosen as the main hazardous pollutants for this investigation. Few methods have been reported for removing these two compounds from water.

Currently, there are different methods for the removal of these contaminants, such as osmosis [13], photocatalytic degradation [14], membrane filtration [15], electrochemical process [16], electrolysis [17], organic processes [18], and adsorption [19,20]. However, techniques other than adsorption are expensive, time-consuming, and complex, with poor recycling efficiency. Adsorption is considered an effective, convenient, and reliable technology because of its simple and straightforward method and operation and high proficiency, even at extremely low concentrations. Several materials have been reported to remove these toxic pollutants, including nanomaterials [21,22], minerals [23], quantum dots [24], metal organic frameworks (MOFs) [25], and covalent organic frameworks (COFs) [26]. Various MOFs employed for wastewater treatment have been reported in the literature. MOFs have a variety of intriguing distinctive traits, including a large surface area, numerous adjustable pores, and tunable surface properties [27]. These features make MOFs promising tools for adsorption, separation, drug delivery, ion exchange, storage, and catalysis [28–30]. Various adsorbents have been reported to remove organic/organometallic molecules. Among these, some materials suffer from several drawbacks, such as complicated or costly synthesis. Like zeolites, which require an inorganic or organic template for their preparation, solvents are the major templating molecules for the synthesis of MOFs [31]. In addition, many materials, such as different aerogels and nanomaterials, have been reported to be less efficient in wastewater treatment compared with MOFs [32–37].

MOF-5 (chemical formula: $Zn_4O_{13}(C_8H_4)_3$), also known as IRMOF-1, a well-known MOF, has vast applications because of its high porosity, unique structure, and thermal stability. It has a three-dimensional structure consisting of Zn_4O clusters and terephthalate ligands [38]. The size and porosity (or pore volume) of an adsorbent determine its adsorption efficiency.

In the present context, we prepared MOF-5 following a procedure reported in the literature with slight modifications and characterized it by FT-IR, powder XRD (PXRD),

FESEM, and EDS analysis. The as-prepared MOF-5 was used to remove safranin O and merbromin from simulated and real wastewater via adsorption. This is the first time MOF-5 has been used to eliminate these two pollutants.

2. Results and Discussion

2.1. Characterizations of the Synthesized MOF-5

2.1.1. PXRD Analysis

Figure 2a shows the PXRD pattern of the synthesized MOF-5. All characteristic peaks appear at 2θ values of 6.86°, 9.71°, 13.6°, 15.3°, 22.5°, 24.9° and 26.3°, which represent the corresponding crystal planes of (200), (220), (400), (420), (442), (711), and (731), respectively. These are very similar to the reported MOF-5 PXRD diffraction data in the literature [39,40] and closely match the file number of the Joint Committee on Powder Diffraction Standards (JCPDS) 36–1451 [41]. From these data, it can be concluded that MOF-5 has a cubic unit cell crystal structure. The sharp and large intense peaks represent the high crystallinity of the adsorbent, which is consistent with the FESEM image (Figure 2c).

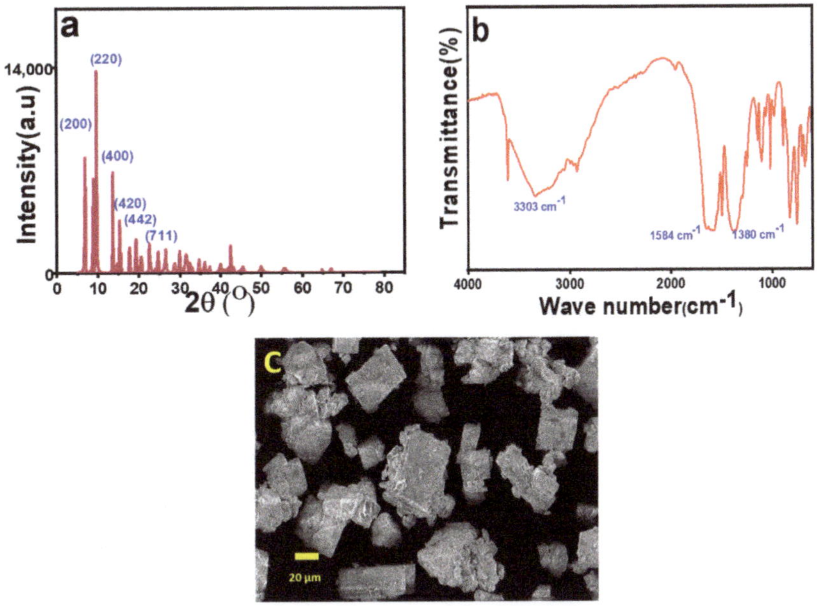

Figure 2. (**a**) PXRD pattern, (**b**) FTIR spectra, and (**c**) FESEM image of MOF-5.

2.1.2. FT-IR Analysis

The skeletal arrangement of MOF-5 was validated using FT-IR spectroscopy. The stretching vibrations of the -COO group in the terephthalic acid linker had symmetric and asymmetric stretching peaks at 1380 cm^{-1} and 1584 cm^{-1} in the FT-IR spectra (Figure 2b). This stretching vibration was different from the infrared spectrum of the terephthalic acid ligand, which appeared at 1660 cm^{-1}. The peaks corresponding to >C=O exhibited a noticeable red shift. This is due to the formation of terephthalate and Zn^{2+} coordination of the ligand -COO, that is, COO-Zn^{2+}. This makes the carbonyl electron distribution relatively uniform and lowers the electron cloud's density [42]. The broad line at 3000–3500 cm^{-1} indicates the presence of water molecules coordinated to the metal center [43]. The sharp peak at 1502 cm^{-1} represents the C=C vibration that is present in the linker [40]. Several small peaks within the ranges of 1250–950 cm^{-1} and 800–650 cm^{-1} represent the in-plane and out-of-plane vibrations of the aromatic C-H bonds [44]. When compared with the data mentioned in the literature, it can be justified that pristine MOF-5 crystals were formed in the present synthesis.

2.1.3. FE-SEM Analysis

The morphology of MOF-5 was investigated by FESEM. In the present work, it was found that the average size of the prepared MOF-5 ranged from 10 to 20 μm (Figure 2c) of the prepared MOF-5. As shown in Figure 2c, all MOF-5 particles were cubic in shape. Mirsoleimani-Azizi et al. synthesized cubic MOF-5 with an average diameter of approximately 550 nm using zinc acetate at room temperature [45]. Cubic MOF-5 with a crystal size of 5–25 μm was synthesized by Son et al. using a rapid sonochemical method [46]. Uniform cubic MOF-5 crystals, 20–25 μm in size, were also prepared through a microwave heating solvothermal route using 1-methyl-2-pyrrolidone as a solvent by SikChoi et al. [47]. Zhao et al. synthesized cubic MOF-5 monocrystals with diameters of 40–60 μm [48].

2.1.4. EDS Analysis

The elemental composition of MOF-5 was analyzed using EDS. The elemental mapping of the uniform distribution of C, O, and Zn in Figure S1 demonstrates how the skeleton function of MOF materials can prevent the aggregation of metal particles and ensure that the appropriate metal elements are equally dispersed, which is very similar to the data reported in the literature [49].

2.2. Effect of pH

Figure 3a shows the variation in percentage removal under different pH conditions for both safranin O and merbromin. The maximum removal efficiency was observed at pH 10 (53.27%) for safranin O and at pH 6 (41.49%) for merbromin. The point of zero charge (PZC) for the MOF-5 was at pH 4.6, which matches well with the reported value [50]. Hence, the surface of MOF-5 becomes negatively charged above pH 4.6 and positively charged below this pH [50]. Safranin O is a cationic azo dye, while merbromin (disodium organomercuric salt) contains a negatively charged exterior owing to the presence of $–O^-$ and $–COO^-$ groups. This information is consistent with that observed in the current study on the effect of pH on pollutant removal. Hence, for safranin O, removal increased as the surface became increasingly negative. For merbromin, an increase was observed as the pH increased from 2 to 6, and above pH 6, the removal efficiency decreased because of the increasing repulsive interaction between the negatively charged merbromin and MOF-5. Significant electrostatic interactions between the adsorbent and adsorbate occurred during adsorption. In addition, there may be a weak noncovalent force of attraction between merbromin and MOF-5, for which it was adsorbed. There may be substantial H bonding between the OH and –COOH groups of merbromin (at pH 6) and the adsorbed –OH on MOF-5 at low concentrations of –OH groups on MOF-5 (at pH-6).

2.3. Effect of Adsorbent Dose

The mass of the adsorbent is also a determining factor, because it provides adsorption sites for adsorbate molecules. Figure 3c shows the variation in the removal percentage with the adsorbent dosage (W). The removal efficiency of the pollutants increased with an increase in the amount of adsorbent (from 250 to 2000 mg/L) and was found to be maximum (95.01% and 90.80% for safranin O and merbromin, respectively) at an adsorbent concentration of 2000 mg/L. This result was obvious, as the increase in the amount of adsorbent provided a greater number of adsorption sites for capturing the adsorbate molecules during the adsorption process, which ultimately enhanced the adsorption [49].

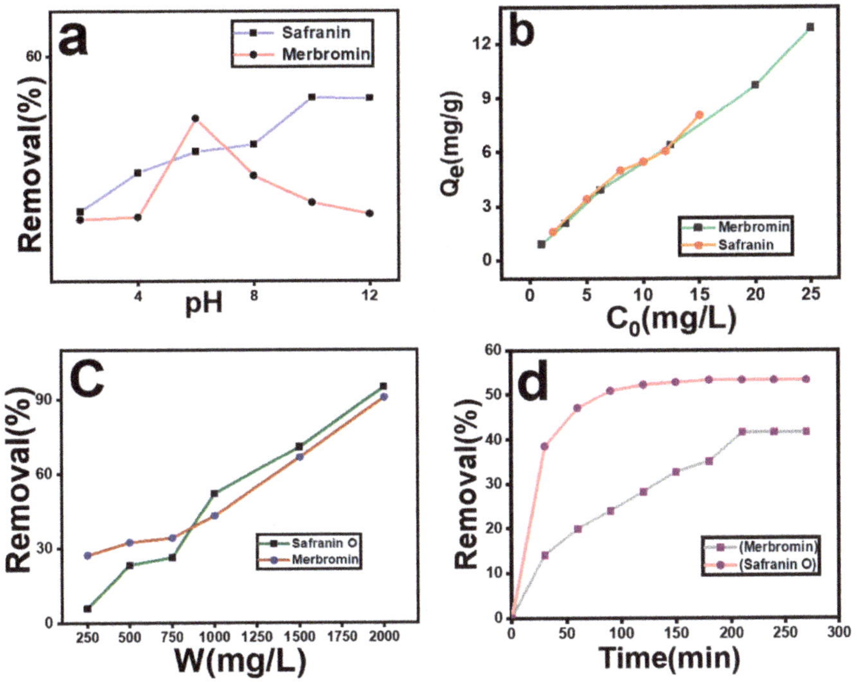

Figure 3. Effect of (**a**) pH, (**b**) initial concentration of contaminant, (**c**) amount of adsorbent, and (**d**) effect of contact time on removal percentage.

2.4. Effect of Contact Time and Adsorption Kinetics

As the adsorption time increased, the pollutant was increasingly adsorbed on MOF-5. The equilibrium times for safranin O and merbromin were 180 min and 210 min, respectively. After equilibrium was reached, the adsorption percentage remained almost unchanged. When equilibrium was reached, the percentage removal was 53.27% and 41.49% for safranin O and merbromin, respectively (Figure 3d). This result was obvious, because when equilibrium was reached, all the adsorption sites were closed, and there were no vacant adsorption sites.

For the kinetic study, the pollutants were adsorbed on MOF-5 for 180 min for safranin O and 210 min for merbromin. The removal percentages in both cases were checked periodically, and the data are plotted in Figure 4a,b. The kinetic data fit well to the linear form of the pseudo-first-order model. All calculations and related figures are presented in Supplementary Materials (Table S1 and Figure S2). Here, for safranin O, as the value of R^2 (=0.98) was close to unity and the value of the experimental adsorption capacity (Q_e(ex) = 7.99 mg/g) was well matched to the calculated value (Q_e(cal) = 6.22 mg/g), it can be concluded that in this case, a pseudo-first order should be the kinetic model. Meanwhile, in the case of merbromin, R^2 was 0.98, and here, Q_e(ex) = 10.37 mg/g, which was matched well with the theoretical one (Q_e(cal) = 9.75 mg/g) for the pseudo-first order. Here, R^2 is the correlation coefficient, as reported in the literature [51]. According to the literature, our experimental data fit well with the pseudo-first-order kinetic model, as evidenced by the highest R^2 value [51,52].

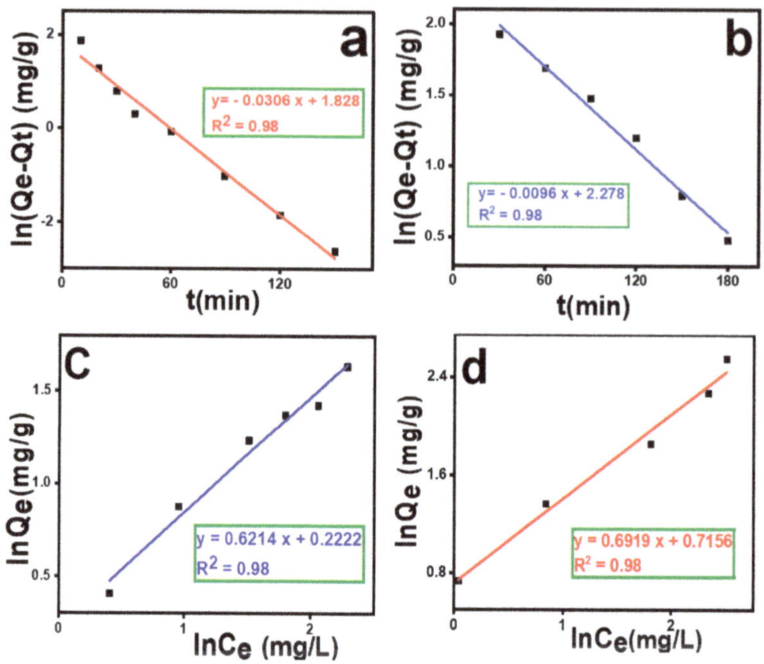

Figure 4. (**a**) Pseudo-first-order plot of safranin O removal, (**b**) pseudo-first-order plot of merbromin removal, (**c**) Freundlich isotherm plot of safranin O adsorption, and (**d**) Freundlich isotherm plot of merbromin adsorption.

2.5. Effect of Adsorbate Dosage and Isotherm Modeling

The variation in the initial concentration (C_o) of waste contaminants is another important factor in the sorption process. The equilibrium adsorption capacity (Q_e) was found to be maximum (8.19 mg/g) at the 15 mg/L concentration and minimum (1.61 mg/g) for the 2 mg/L solution of safranin O. Meanwhile, in the case of merbromin, the equilibrium adsorption capacity (Q_e) was found to be maximum (12.22 mg/g) at a higher concentration (25 mg/L) and minimum (0.09 mg/g) at a lower concentration (1 mg/L) (Figure 3b). This occurs because at higher concentrations of pollutant molecules, there is a higher probability of collisions with adsorption sites on the adsorbent surface. Again, the mass transfer improves during adsorption, which can decrease the mass transfer resistance and hence increase the adsorption capacity [53,54].

Isotherm modeling is important for this analysis, as it provides information regarding the isotherm adsorption behavior for the removal process of these contaminants. Five isotherm models (Langmuir isotherm, Freundlich isotherm, Temkin isotherm, Elovich isotherm, and Dubinin–Radushkevich isotherm models) were investigated to determine the adsorption process. The results of the experiment are shown in Figures 4c,d, S3 and S4 and Table S2. From the data, it can be concluded that the best-fitting model for both cases was the Freundlich isotherm. Thus, the adsorption was multilayered, with K_F = 1.24 [mg. g^{-1} (mg $L^{-1})^{-0.6214}$] (for safranin O) and 2.04 [mg. g^{-1} (mg $L^{-1})^{-0.6919}$] (for merbromin), and n values were 1.60 (for safranin O) and 1.44 (for merbromin). Linear fitting was performed using the R^2.

Different Q_{max} values for dye removal by MOF-5 are reported in Table 1 (if we assume that the removal follows the Langmuir isotherm model, then only the Q_{max} values were 8.19 mg/g and 22.22 mg/g for safranin O and merbromin, respectively (Table S2)). From the table, it can be concluded that our prepared MOF-5 also works as well as the other reported MOF-5s, and that it is even superior to some other MOF-5s.

Table 1. Comparison of Q_{max} for different dye removals by MOF-5.

Adsorbate	Type of Dye	Q_{max} (mg/g)	References
Aniline blue	Cationic	55.34	[49]
Methylene blue	Cationic	51.81	[55]
Malachite green	Cationic	50.69	[56]
Methylene blue	Cationic	2.632	[49]
Safranin O	Cationic	8.19	This study
Congo Red	Anionic	769.23	[57]
Orange II	Anionic	10.01	[49]
Merbromin	Anionic	22.22	This study

Overall, it can be concluded that MOF-5 had certain advantages over other different adsorbents. It has also been found that MOFs are more or similarly efficient in comparison to other kinds of adsorbents (e.g., nanomaterials, bio adsorbents, graphene oxide, carbon nanotubes, and aerogels) that are used for wastewater treatment. Comparisons between the Q_{max} values of different adsorbents for the removal of different dyes are reported in Table 2 (to the best of our knowledge, merbromin removal has not yet been reported).

Table 2. Comparison of Q_{max} for different dyes with MOFs and other adsorbents.

Adsorbent	Adsorbate	Q_{max} (mg/g)	References
Coconut coir	Methylene blue	15.59	[32]
Egg shell	Methylene blue	16.43	[33]
Activated carbon from waste biomass	Methylene blue	10.21	[34]
ZnO hydrid beads	Basic blue 41	1.0–8.0	[35]
Polylactide/spent grain	Malachite green	1.48	[36]
ZnO nanoparticles	Methylene blue	0.3428	[37]
Coir pith carbon	Methylene blue	5.87	[58]
ZnO@ananas comosus waste biomass	Celestine blue	6.52	[37]
Ananas comosus waste biomass	Celestine blue	5.42	[37]
Brewery spent grain	Malachite green	2.55	[36]
Poultry feathers	Malachite green	3.55	[36]
Ananas comosus waste biomass	Celestine blue	5.42	[37]
MOF-5	Merbromin	22.22	This study
MOF-5	Safranin O	8.19	This study

2.6. Thermodynamics of Adsorption

The adsorption removal efficiency with respect to temperature is shown in Figure 5a, and the data are represented in Table S3. The removal efficiency increased with increasing temperature in both cases. The data were plotted as [ln $(1/C_e)$ vs. temperature $(1/T)$] (Figure 5b), and the thermodynamics of adsorption were modeled using linear regression curves. The R^2 value was higher than 0.98, which proved that the curves were well fitted. ΔH values were calculated to be 18.77 kJ/mol and 11.16 kJ/mol for safranin O and merbromin, respectively, from Figure 5b (overall results are displayed in Table S3). A positive ΔH value indicates that both adsorptions are endothermic in nature, and a negative ΔG value implies that the adsorption is spontaneous [59]. The ΔG value is important because it indicates whether the adsorption is physisorption or chemisorption. If the ΔG value is negative and ranging from 0 to 20 kJ/mol, then it is physisorption, and if it is negative within the range of 80 kJ/mol to 400 kJ/mol, then it is chemisorption [60]; however, the ΔG° value varied from −3.8 to −4.2 kJ/mol for safranin O and from −3.4 to −3.8 kJ/mol for merbromin, and in both cases, the negative ΔG value was lower than 20 kJ/mol; hence, the adsorption process was physisorption [49,61]. Here, Δ*S* was highly positive, indicating that during the adsorption process, the degrees of freedom increased at the solid–liquid interface [45].

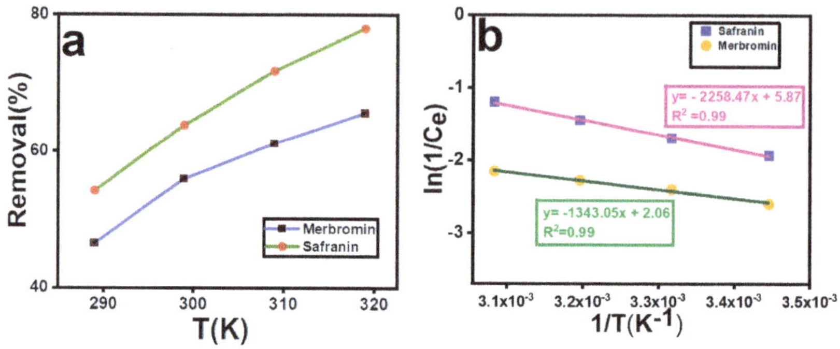

Figure 5. (**a**) Effect of temperature on removal percentage, (**b**) plot of $\ln(1/C_e)$ vs. (1/T).

2.7. Effect of Natural Ions and Molecules Present in Real Water

To determine the impact of other metal ions, anions, small molecules, and other microorganisms on the adsorption behavior of the two organic pollutants, an adsorption study in lake water was performed. The removal of various ions (Na^+, K^+, F^-, Cl^-, SO_4^{2-}, PO_4^{3-}, Mg^{2+}, and Ca^{2+}) from contaminated real water was performed using the same concentrations of the two pollutants. At room temperature and neutral pH, the adsorption percentages of safranin O and merbromin were 34.00% and 26.28%, respectively. These values are lower than those in the simulated wastewater. These changes were due to several ions and other molecules being present in the real wastewater sample, which competed with these pollutants and reduced their adsorption capacity. Table S4 presents their amounts, and a comparison of the percentages of adsorption of the pollutants in the simulated wastewater and lake water is presented in Figure 6a.

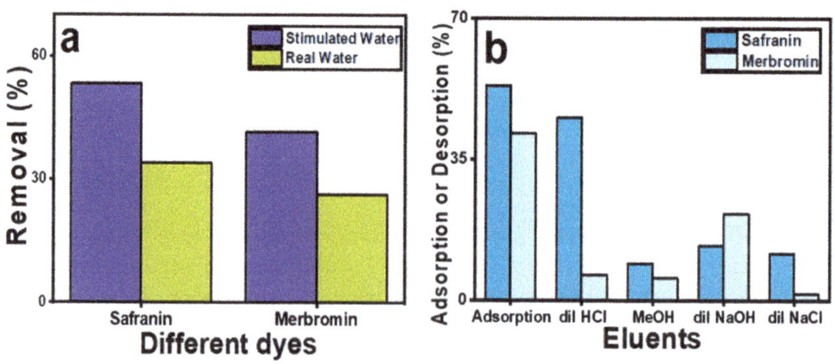

Figure 6. Comparison of removal percentage in (**a**) simulated wastewater and surface water containing various ions; (**b**) desorption of the dyes in different eluents.

2.8. Regeneration of Adsorbent

Figure 6b shows the desorption percentages of both dyes in the different solvents after a single desorption cycle. It was found that after 180 min, the maximum desorption percentage for safranin O was 45.4% in HCl, for merbromin, it was 21.6% in NaOH, and it then remained constant after further exposure. The desorption percentages of safranin O were found to be 13.6% for NaOH, 11.7% for NaCl, and 9.2% for methanol. This is acceptable, as safranin O is a cationic dye, and H_3O^+ can effectively replace safranin O, because it can form a stronger interaction with MOF-5 than with safranin O. Furthermore, NaOH and NaCl both form the same cation Na^+, so their desorption ability is comparable because of their poor charge density due to their larger size compared to H_3O^+.

However, for merbromin, the desorption percentages were 6.3% in HCl, 1.7% in NaCl, and 5.6% in methanol. In the presence of NaOH, owing to the high pH of the eluent, the surface of MOF-5 became highly negatively charged (PZC for MOF-5 was 4.6), along with the formation of more OH^- ions within the solution. Therefore, OH^- ions can effectively replace the anionic merbromin from the MOF-5 surface compared to other conditions. As the desorption percentage was lower than 50%, further cycles were not performed.

2.9. Plausible Mechanism of Adsorption

MOFs can have both positive and negative charges on their surface depending on the pH, while safranin O is a positively charged dye, and merbromin is anionic in nature [62]. If MOFs have a positive charge, they can attract negative or anionic dyes, and if MOFs are negatively charged, they can attract positively charged dyes electrostatically. Therefore, electrostatic attraction is the predominant factor in the adsorption of MOFs [63]. Dyes can also be adsorbed via π-π interaction [64], hydrogen bonding [65–67], ion exchange [68,69], Lewis acid–base interactions [70–72], etc.

For electrostatic interactions, the MOF surface and the adsorbate should have opposite charges, so that they can easily be attracted to each other and hence adsorbed through a pure Coulombic attraction force. In the case of π–π-type interactions, noncovalent interactions occur between the two aromatic rings. Hazrati and Safari [73] reported that the sorption of reactive black 5 on a Cd-based MOF (TMU-8) is due to π–π interaction between the aromatic ring of reactive black 5 and the framework of the MOF. Again, an exchange of ligands or ions between the two intricate structures is found for the sorption of dyes onto MOFs via the ion exchange process. According to Yao et al. [68], adsorption occurs via ion exchange between the $[(CH_3)_2NH_2]^+$ ions that are present in JLU-Liu 39 and cationic dyes (methyl violet, methylene blue, and rhodamine B) [68]. According to Zhao et al., the presence of Ni^{2+} as a Lewis acid enhances the removal of Congo red onto GO/MOFs through a Lewis acid–base interaction [72]. There is also a mechanism called adsorption through hydrogen bonding, which has rarely been reported.

According to the adsorption kinetics in the present context, MOF-5 is more useful and has a better adsorption capacity for cationic dyes (safranin O) than for merbromin. This may be explained by the electrostatic interactions between the negatively charged MOF-5 surface and the cationic dye (Figure 7). The carboxylate moiety of MOF-5 forms a negatively charged framework. At pH 10, the framework becomes negatively charged (PZC for MOF-5 was 4.6), thus enhancing the adsorption of safranin O.

In contrast, merbromin is electronically anionic, and at a lower pH, it becomes protonated and remains mostly neutral. Thus, at a lower pH, there might be a weak π-π-type interaction, but as the pH of the medium increases, there will be electrostatic repulsion between the merbromin molecule and the MOF-5 surface. Furthermore, there may be a major chance in the interaction between the aromatic ring of the ligand that is present in the MOF and the lone pairs of oxygen that are present in the merbromin dye at a lower pH. This π-π-type interaction was reported by Elsherbiny et al. [49] and was the main cause for the adsorption of merbromin onto MOF-5. As the π-π-type interaction is weaker than the pure electrostatic interaction, the removal percentage of merbromin is comparatively lower than that of safranin O.

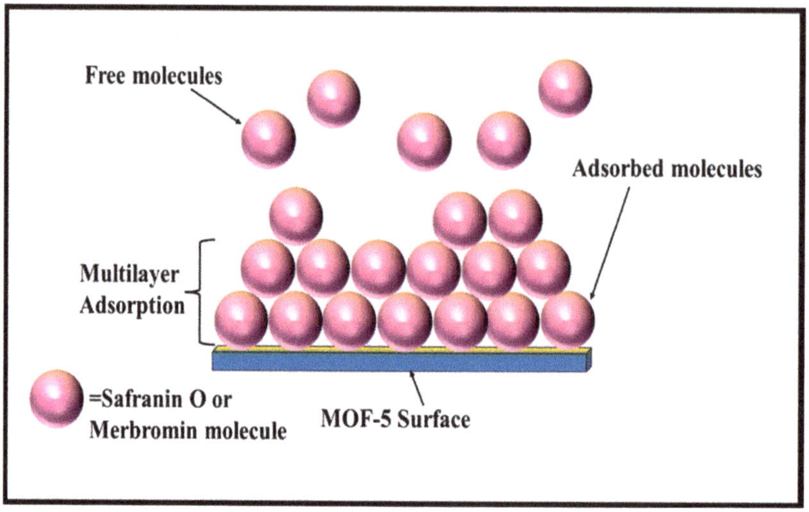

Figure 7. Plausible mechanistic pathway for multilayer adsorption of safranin O and merbromin on MOF-5.

3. Materials and Methods

3.1. Chemicals and Reagents

Safranin was purchased from Sigma-Aldrich, India, merbromin from Loba Chemie, Mumbai, India, and other common reagents were purchased from a local chemical company. All chemicals used in this study were of analytical grade and used without further purification. All solutions used in the adsorption experiments were prepared using Milli-Q water. The adsorption of these pollutants onto MOF-5 was also evaluated in a real wastewater sample, collected from Kochbihar Lake (Sagar Dighi), Cooch Behar, West Bengal, India, on 12 June 2023.

3.2. Preparation of MOF-5

MOF-5 was synthesized following a previously reported procedure [74] with some modifications. First, terephthalic acid (0.033 g, 0.2 mmol) was dissolved in 20 mL of DMF in a conical flask, and then, the zinc nitrate tetrahydrate (0.156 g, 0.6 mmol) solution (in 20 mL of DMF) was added with continuous stirring. The reaction mixture was heated in an oil bath at 110 °C for 24 h, cooled to room temperature, repeatedly washed with DMF followed by anhydrous chloroform, soaked for 24 h in chloroform, filtered, and vacuum-dried for 24 h. MOF-5 was then kept under vacuum for 2 h, while being activated at 105 °C.

3.3. Characterization Techniques

The common instruments used to perform the experiments were a Remi R-8C centrifuge, Remi orbital shaker (Model RS-36BL, Remi, Mumbai, India), Fisher Scientific Accumet pH meter (Model AB 250, Fisher Scientific, Loughborough LE11 5RG, UK), and Milli-Q Plant from Labconco Water Pro/Ro, Labconco, Kansas City, MO, USA. Skeletal analysis of MOF-5 was performed using Fourier transform infrared (FTIR) spectroscopy on a Benchtop Labtronics LT-4100, Labtronics, Welwyn Garden City AL7 1TW, UK, and crystallinity was investigated by powder X-ray diffraction (PXRD) on a Thermo Scientific ARL Equinox 1000, ThermoFisher Scientific, Waltham, MA, USA. Morphology, size, and elemental analyses were performed using field-emission scanning electron microscopy (FESEM) and energy-dispersive spectroscopy (EDS) (Zeiss EVO 18 from IIT Palakkad, Kerala, India). The experimental analysis was performed using a UV-vis spectrophotometer (Thermo Scientific Evolution 201, Fisher Scientific, Loughborough LE11 5RG, UK).

3.4. Adsorption Isotherm Model

Adsorption isotherm modeling is an important parameter. Numerous adsorption isotherm models have been developed to fit the experimental adsorption equilibrium data. This study utilized commonly used solid–liquid adsorption isotherm models. The different adsorption isotherm models with their equations and other statistical parameters are plotted in Table S2 [61,75–77].

3.5. Adsorption Thermodynamics

For the thermodynamic study of adsorption, different authors have reported different techniques for calculating adsorption thermodynamic parameters [78–81]. In this regard, as adsorption follows the Freundlich model, we consider the equations reported in the literature [76,82].

$$\ln\frac{1}{C_e} = \ln K_o - \frac{\Delta H}{RT} \tag{1}$$

$$\Delta G^o = -nRT \tag{2}$$

$$\Delta S^o = \frac{\Delta H^o - \Delta G^o}{T} \tag{3}$$

where C_e is the equilibrium adsorbate concentration (mg/L) in the solution, n is the fitting constant of the Freundlich exponent, K_F is the Freundlich empirical constant. R is the universal gas constant (8.314 J/(mol K)) and T is the temperature (K). From the slope of the plot of ln $1/C_e$ versus 1/T, the enthalpy change in adsorption (ΔH°) was calculated.

4. Experimental

4.1. Batch Adsorption Experiments

A comparative study of the adsorption behavior of MOF-5 towards safranin O and merbromin was performed using the batch method. To study the effect of the initial adsorbate concentration, different concentrations of safranin O (2–15 mg/L) and merbromin (1–25 mg/L) were prepared from 100 mg/L stock solutions of the two pollutants. The removal percentages of pollutants were also studied using different amounts of adsorbent, ranging from 250 to 2000 mg/L. Adsorption was studied in the pH range of 2–12 to determine the effect of pH on removal. The pH of the solution was adjusted using 0.1 M NaOH and 0.1 M HCl. Adsorption experiments were conducted at different temperatures (289, 299, 309, and 319 K) to determine the thermodynamic parameters of the process, such as the Gibbs free energy change (ΔG), enthalpy change (ΔH), and entropy change (ΔS).

Except where otherwise stated, the sorption experiment was carried out at room temperature (r.t.) using 15 mg/L of safranin O at pH 10 and 25 mg/L of merbromin at pH 6; the volume of the solution was 10 mL. In all the cases, the amount of MOF-5 was 10 mg, which remained the same unless otherwise stated. Safranin O and merbromin were mixed with MOF-5 separately for 180 min and 210 min, respectively, and shaken at 250 rpm in an orbital shaker from Remi (model number: RS-36BL, Remi, Mumbai, India). Aliquots were periodically collected, centrifuged, and monitored using a UV-Vis spectrophotometer (519 nm for safranin O and 505 nm for merbromin) to evaluate the progress of the reaction. Before the UV-Vis spectral scanning for the determination of the pollutant concentrations, centrifugation was performed every time on a Remi centrifugation (Remi R-8C, Remi, Mumbai, India) at 1200 rpm for 10 min to extract MOF-5 from the solution.

The adsorption percentage and capacity (q_t) were calculated using Equations (4) and (5), respectively.

$$\text{Adsorption } (\%) = \frac{(C_0 - C_t)}{C_0} \times 100 \tag{4}$$

$$q_t = \frac{(C_0 - C_t)}{m} \times V \tag{5}$$

where C_o and C_t are the dye concentrations (mg/L) initially and after adsorption, respectively, and V and m are the volume (L) of the solution and mass (g) of MOF-5, respectively.

Because the adsorption process was performed in an aqueous medium, the aqueous stability of MOF-5 was an important issue. In many studies, it was mentioned that MOF-5 showed considerable stability in aqueous media even after 24 h, and as the equilibrium of adsorption was reached within 24 h in the present study, in the presence of water, the stability and the adsorption procedure were not hampered. For example, Elsherbiny et al. [49] concluded from the TGA data that Zn-BDC (MOF-5) is considerably stable upto 425 °C, which could be advantageous for the removal of pollutants at high temperatures. They also performed a test of stability in water, and the results were satisfactory. Therefore, MOF-5 could be used to treat wastewater. There are other examples too, such as Mohammadi et al. [56] used MOF-5 to remove malachite green from simulated wastewater.

4.2. Regeneration of Adsorbent

In addition to adsorption, regeneration (or desorption) is an important mechanism by which MOF can be recovered and reused. This was studied after the adsorption process for both safranin O and merbromin was completed. In this regard, MOF-5 was initially well saturated with the adsorbate solutions (15 mg/L at pH 10 for safranin O and 25 mg/L at pH 6 for merbromin; volume in each case = 10 mL); thereafter, centrifugation and decantation were performed, followed by washing with Milli-Q water. For the regeneration step, the adsorbate-loaded MOF-5 was then mixed with 0.01 M HCl, 0.01 M NaOH, 0.01 M NaCl, and 99.8% MeOH separately and shaken till equilibrium [83].

The percentage of desorption was determined by the following equation (Equation (6)):

$$\text{Desorption }(\%) = \frac{\text{amount of dye desorbed}}{\text{amount of dye adsorbed}} \times 100 \quad (6)$$

5. Conclusions

In general, MOF-5 was effectively synthesized and characterized by FT-IR, PXRD, FESEM, and EDS using the reported data. The MOF-5 particles were cubic. Two organic pollutants, safranin O and merbromin, were removed successfully using MOF-5 from the simulated wastewater. The pollutant removal followed pseudo-first-order kinetics and the Freundlich adsorption isotherm model. Moreover, thermodynamic parameters demonstrated that adsorption occurs naturally as physisorption and is thermodynamically favorable ($\Delta G = -$ ve for both safranin O and merbromin). A reduction in the percentage of adsorption occurred when real wastewater containing various ions (Na^+, K^+, F^-, Cl^-, $SO_4{}^{2-}$, $PO_4{}^{3-}$, Mg^{2+}, and Ca^{2+}) was used. Safranin O and merbromin are widely used in biological staining and pharmaceuticals despite their hazardous effects. We hope that this removal study will be helpful in controlling water pollution that is caused by industrial and pharmaceutical issues as much as possible by adding this new method for the improvement of the environment.

Supplementary Materials: The following supporting information can be downloaded at https://www.mdpi.com/article/10.3390/molecules29040886/s1, Figure S1: (a). EDS spectra of as synthesized MOF-5 and (b). elemental mapping of MOF-5; Figure S2: (a) pseudo second order plot of merbromin removal, (b) pseudo second order plot of safranin O removal, (c) intra-particle diffusion of merbromin removal, (d) intra particle diffusion of safranin O removal; Table S1: Statistical data for different Adsorption Kinetics; Figure S3: Different Adsorption Isotherms for Merbromin adsorption onto MOF-5 (a) Langmuir isotherm, (b) Temkin isotherm, (c) Elovich isotherm, (d) Dubinin-Radushkevich isotherm; Figure S4: Different Adsorption Isotherms for Safranin O adsorption onto MOF-5 (a) Langmuir isotherm, (b) Temkin isotherm, (c) Elovich isotherm, (d) Dubinin-Radushkevich isotherm; Table S2: Statistical data for different adsorption isotherm; Table S3: Thermodynamic parameters for dye adsorption onto MOF-5; Table S4: Various parameters in Kochbihar lake water.

Author Contributions: Conceptualization, G.B.; Data curation, C.D., M.P. and J.I.; Formal analysis, N.R., C.D. and M.P.; Funding acquisition, J.I. and G.B.; Investigation, N.R., C.D. and M.P.; Methodology, N.R., C.D. and G.B.; Supervision, J.I. and G.B.; Validation, C.D.; Writing—original draft, N.R.; Writing—review and editing, J.I. and G.B. All authors have read and agreed to the published version of the manuscript.

Funding: This work was funded by BK21 FOUR (Fostering Outstanding Universities for Research) (No. 5199991614564) and by the Soonchunhyang University Research Fund.

Institutional Review Board Statement: Not applicable for this study.

Informed Consent Statement: Not applicable for this study.

Data Availability Statement: The data that support the findings of this study are available from the corresponding author upon reasonable request.

Acknowledgments: N.R., C.D. and G.B. are thankful to Department of Chemistry, Cooch Behar Panchanan Barma University, for providing the necessary funding and instrumental facility. The authors would like to thank the central instrumental facility of IIT Palakkad, Kerala, for the FESEM and EDS data.

Conflicts of Interest: The authors declare no competing financial interests.

References

1. Maheshwari, K.; Solanki, Y.S.; Ridoy, M.S.H.; Agarwal, M.; Dohare, R.; Gupta, R. Ultrasonic Treatment of Textile Dye Effluent Utilizing Microwave-assisted Activated Carbon. *Environ. Prog. Sustain. Energy* **2020**, *39*, e13410. [CrossRef]
2. Arif, A. Water Pollution and Industries. *Pure Appl. Biol.* **2020**, *9*, 2214–2224. [CrossRef]
3. Banat, I.M.; Nigam, P.; Singh, D.; Marchant, R. Microbial Decolorization of Textile-Dyecontaining Effluents: A Review. *Bioresour. Technol.* **1996**, *58*, 217–227. [CrossRef]
4. Liu, Q. Pollution and Treatment of Dye Waste-Water. *IOP Conf. Ser. Earth Environ. Sci.* **2020**, *514*, 052001. [CrossRef]
5. Salleh, M.A.M.; Mahmoud, D.K.; Karim, W.A.W.A.; Idris, A. Cationic and Anionic Dye Adsorption by Agricultural Solid Wastes: A Comprehensive Review. *Desalination* **2011**, *280*, 1–13. [CrossRef]
6. Winckler, J. Vitalfärbung von Lysosomen und anderen Zellorganellen der Ratte mit Neutralrot Vital Staining of Lysosomes and Other Cell Organelles of the Rat with Neutral Red. *Prog. Histochem. Cytochem.* **1974**, *6*, 1–89. [CrossRef] [PubMed]
7. Carvalho Pinheiro, N.S.; Perez-Lopez, O.W.; Gutterres, M. Solid Leather Wastes as Adsorbents for Cationic and Anionic Dye Removal. *Environ. Technol.* **2022**, *43*, 1285–1293. [CrossRef]
8. Samal, K.; Mahapatra, S.; Hibzur Ali, M. Pharmaceutical Wastewater as Emerging Contaminants (EC): Treatment Technologies, Impact on Environment and Human Health. *Energy Nexus* **2022**, *6*, 100076. [CrossRef]
9. Wall, A.; Board, T. Chemical Basis for the Histological Use of Safranin O in the Study of Articular Cartilage. In *Classic Papers in Orthopaedics*; Banaszkiewicz, P.A., Kader, D.F., Eds.; Springer: London, UK, 2014; pp. 433–435. ISBN 978-1-4471-5450-1.
10. Sharifpour, E.; Ghaedi, M.; Nasiri Azad, F.; Dashtian, K.; Hadadi, H.; Purkait, M.K. Zinc Oxide Nanorod-loaded Activated Carbon for Ultrasound-assisted Adsorption of Safranin O: Central Composite Design and Genetic Algorithm Optimization. *Appl. Organomet. Chem.* **2018**, *32*, e4099. [CrossRef]
11. Aronson, J.K. Mercury and Mercurial Salts. In *Meyler's Side Effects of Drugs*; Elsevier: Amsterdam, The Netherlands, 2016; pp. 844–852, ISBN 978-0-444-53716-4.
12. Hill, S.L.; Thomas, S.H.L. Clinical Toxicology of Newer Recreational Drugs. *Clin. Toxicol.* **2011**, *49*, 705–719. [CrossRef]
13. Li, M.; Wang, X.; Porter, C.J.; Cheng, W.; Zhang, X.; Wang, L.; Elimelech, M. Concentration and Recovery of Dyes from Textile Wastewater Using a Self-Standing, Support-Free Forward Osmosis Membrane. *Environ. Sci. Technol.* **2019**, *53*, 3078–3086. [CrossRef]
14. Kumari, H.; Sonia; Suman; Ranga, R.; Chahal, S.; Devi, S.; Sharma, S.; Kumar, S.; Kumar, P.; Kumar, S.; et al. A Review on Photocatalysis Used For Wastewater Treatment: Dye Degradation. *Water Air Soil Pollut.* **2023**, *234*, 349. [CrossRef] [PubMed]
15. Guo, J.; Jiang, D.; Wu, Y.; Zhou, P.; Lan, Y. Degradation of Methyl Orange by Zn(0) Assisted with Silica Gel. *J. Hazard. Mater.* **2011**, *194*, 290–296. [CrossRef] [PubMed]
16. Wei, K.; Cui, T.; Huang, F.; Zhang, Y.; Han, W. Membrane Separation Coupled with Electrochemical Advanced Oxidation Processes for Organic Wastewater Treatment: A Short Review. *Membranes* **2020**, *10*, 337. [CrossRef] [PubMed]
17. Palanisamy, S.; Nachimuthu, P.; Awasthi, M.K.; Ravindran, B.; Chang, S.W.; Palanichamy, M.; Nguyen, D.D. Application of Electrochemical Treatment for the Removal of Triazine Dye Using Aluminium Electrodes. *J. Water Supply Res. Technol. Aqua* **2020**, *69*, 345–354. [CrossRef]
18. Bhatia, D.; Sharma, N.R.; Singh, J.; Kanwar, R.S. Biological Methods for Textile Dye Removal from Wastewater: A Review. *Crit. Rev. Environ. Sci. Technol.* **2017**, *47*, 1836–1876. [CrossRef]
19. Yagub, M.T.; Sen, T.K.; Afroze, S.; Ang, H.M. Dye and Its Removal from Aqueous Solution by Adsorption: A Review. *Adv. Colloid Interface Sci.* **2014**, *209*, 172–184. [CrossRef] [PubMed]

20. Das, C.; Ghosh, N.N.; Pulhani, V.; Biswas, G.; Singhal, P. Bio-Functionalized Magnetic Nanoparticles for Cost-Effective Adsorptio of U(VI): Experimental and Theoretical Investigation. *RSC Adv.* **2023**, *13*, 15015–15023. [CrossRef]
21. Xu, P.; Zeng, G.M.; Huang, D.L.; Feng, C.L.; Hu, S.; Zhao, M.H.; Lai, C.; Wei, Z.; Huang, C.; Xie, G.X.; et al. Use of Iron Oxid Nanomaterials in Wastewater Treatment: A Review. *Sci. Total Environ.* **2012**, *424*, 1–10. [CrossRef]
22. Das, C.; Sen, S.; Singh, T.; Ghosh, T.; Paul, S.S.; Kim, T.W.; Jeon, S.; Maiti, D.K.; Im, J.; Biswas, G. Green Synthesis, Characterizatio and Application of Natural Product Coated Magnetite Nanoparticles for Wastewater Treatment. *Nanomaterials* **2020**, *10*, 161 [CrossRef]
23. Li, Y.; Li, L.; Yu, J. Applications of Zeolites in Sustainable Chemistry. *Chem* **2017**, *3*, 928–949. [CrossRef]
24. Rani, U.A.; Ng, L.Y.; Ng, C.Y.; Mahmoudi, E. A Review of Carbon Quantum Dots and Their Applications in Wastewater Treatmen *Adv. Colloid Interface Sci.* **2020**, *278*, 102124. [CrossRef] [PubMed]
25. Liu, X.; Shan, Y.; Zhang, S.; Kong, Q.; Pang, H. Application of Metal Organic Framework in Wastewater Treatment. *Green Energ Environ.* **2023**, *8*, 698–721. [CrossRef]
26. Aslam, A.A.; Irshad, A.; Nazir, M.S.; Atif, M. A Review on Covalent Organic Frameworks as Adsorbents for Organic Pollutant *J. Clean. Prod.* **2023**, *400*, 136737. [CrossRef]
27. Stock, N.; Biswas, S. Synthesis of Metal-Organic Frameworks (MOFs): Routes to Various MOF Topologies, Morphologies, an Composites. *Chem. Rev.* **2012**, *112*, 933–969. [CrossRef]
28. Tan, Y.-X.; He, Y.-P.; Wang, M.; Zhang, J. A Water-Stable Zeolite-like Metal–Organic Framework for Selective Separation o Organic Dyes. *RSC Adv.* **2014**, *4*, 1480–1483. [CrossRef]
29. Chen, Q.; He, Q.; Lv, M.; Xu, Y.; Yang, H.; Liu, X.; Wei, F. Selective Adsorption of Cationic Dyes by UiO-66-NH2. *Appl. Surf. Sc* **2015**, *327*, 77–85. [CrossRef]
30. Wang, C.; Liu, X.; Keser Demir, N.; Chen, J.P.; Li, K. Applications of Water Stable Metal–Organic Frameworks. *Chem. Soc. Re* **2016**, *45*, 5107–5134. [CrossRef] [PubMed]
31. Khan, N.A.; Hasan, Z.; Jhung, S.H. Adsorptive Removal of Hazardous Materials Using Metal-Organic Frameworks (MOFs): Review. *J. Hazard. Mater.* **2013**, *244–245*, 444–456. [CrossRef] [PubMed]
32. Sharma, Y.C.; Upadhyay, S.N. Removal of a Cationic Dye from Wastewaters by Adsorption on Activated Carbon Developed fro Coconut Coir. *Energy Fuels* **2009**, *23*, 2983–2988. [CrossRef]
33. Tsai, W.T.; Yang, J.M.; Lai, C.W.; Cheng, Y.H.; Lin, C.C.; Yeh, C.W. Characterization and Adsorption Properties of Eggshells an Eggshell Membrane. *Bioresour. Technol.* **2006**, *97*, 488–493. [CrossRef]
34. Karagoz, S.; Tay, T.; Ucar, S.; Erdem, M. Activated Carbons from Waste Biomass by Sulfuric Acid Activation and Their Use o Methylene Blue Adsorption. *Bioresour. Technol.* **2008**, *99*, 6214–6222. [CrossRef]
35. Shokry Hassan, H.; Elkady, M.F.; El-Shazly, A.H.; Bamufleh, H.S. Formulation of Synthesized Zinc Oxide Nanopowder int Hybrid Beads for Dye Separation. *J. Nanomater.* **2014**, *2014*, 967492. [CrossRef]
36. Muinde, V.M.; Onyari, J.M.; Wamalwa, B.; Wabomba, J.N. Adsorption of Malachite Green Dye from Aqueous Solutions Usin Mesoporous Chitosan–Zinc Oxide Composite Material. *Environ. Chem. Ecotoxicol.* **2020**, *2*, 115–125. [CrossRef]
37. Akpomie, K.G.; Conradie, J. Synthesis, Characterization, and Regeneration of an Inorganic–Organic Nanocomposit (ZnO@biomass) and Its Application in the Capture of Cationic Dye. *Sci. Rep.* **2020**, *10*, 14441. [CrossRef]
38. Gangu, K.K.; Maddila, S.; Jonnalagadda, S.B. The Pioneering Role of Metal–Organic Framework-5 in Ever-Growing Contemporar Applications—A Review. *RSC Adv.* **2022**, *12*, 14282–14298. [CrossRef]
39. Greer, H.F.; Liu, Y.; Greenaway, A.; Wright, P.A.; Zhou, W. Synthesis and Formation Mechanism of Textured MOF-5. *Cryst. Growt Des.* **2016**, *16*, 2104–2111. [CrossRef]
40. Biserčić, M.S.; Marjanović, B.; Vasiljević, B.N.; Mentus, S.; Zasońska, B.A.; Ćirić-Marjanović, G. The Quest for Optimal Wate Quantity in the Synthesis of Metal-Organic Framework MOF-5. *Microporous Mesoporous Mater.* **2019**, *278*, 23–29. [CrossRef]
41. Barthwal, S.; Jeon, Y.; Lim, S.-H. Superhydrophobic Sponge Decorated with Hydrophobic MOF-5 Nanocoating for Efficien Oil-Water Separation and Antibacterial Applications. *Sustain. Mater. Technol.* **2022**, *33*, e00492. [CrossRef]
42. Wang, S.; Xie, X.; Xia, W.; Cui, J.; Zhang, S.; Du, X. Study on the Structure Activity Relationship of the Crystal MOF-5 Synthesis Thermal Stability and N2 Adsorption Property. *High Temp. Mater. Process.* **2020**, *39*, 171–177. [CrossRef]
43. Wang, S.; Ye, B.; An, C.; Wang, J.; Li, Q.; Guo, H.; Zhang, J. Exploring the Coordination Effect of GO@MOF-5 as Catalyst o Thermal Decomposition of Ammonium Perchlorate. *Nanoscale Res. Lett.* **2019**, *14*, 345. [CrossRef] [PubMed]
44. Iswarya, N.; Kumar, M.G.; Rajan, K.S.; Rayappan, J.B.B. Metal Organic Framework (MOF-5) For Sensing of Volatile Organi Compounds. *J. Appl. Sci.* **2012**, *12*, 1681–1685. [CrossRef]
45. Mirsoleimani-azizi, S.M.; Setoodeh, P.; Zeinali, S.; Rahimpour, M.R. Tetracycline Antibiotic Removal from Aqueous Solutions b MOF-5: Adsorption Isotherm, Kinetic and Thermodynamic Studies. *J. Environ. Chem. Eng.* **2018**, *6*, 6118–6130. [CrossRef]
46. Son, W.-J.; Kim, J.; Kim, J.; Ahn, W.-S. Sonochemical Synthesis of MOF-5. *Chem. Commun.* **2008**, *47*, 6336–6338. [CrossRef] [PubMed]
47. Choi, J.-S.; Son, W.-J.; Kim, J.; Ahn, W.-S. Metal–Organic Framework MOF-5 Prepared by Microwave Heating: Factors to B Considered. *Microporous Mesoporous Mater.* **2008**, *116*, 727–731. [CrossRef]
48. Zhao, Z.; Li, Z.; Lin, Y.S. Adsorption and Diffusion of Carbon Dioxide on Metal−Organic Framework (MOF-5). *Ind. Eng. Chem Res.* **2009**, *48*, 10015–10020. [CrossRef]

49. Elsherbiny, A.S.; Rady, A.; Abdelhameed, R.M.; Gemeay, A.H. Efficiency and Selectivity of Cost-Effective Zn-MOF for Dye Removal, Kinetic and Thermodynamic Approach. *Environ. Sci. Pollut. Res.* **2023**, *30*, 106860–106875. [CrossRef]
50. Wu, Y.; Pang, H.; Yao, W.; Wang, X.; Yu, S.; Yu, Z.; Wang, X. Synthesis of Rod-like Metal-Organic Framework (MOF-5) Nanomaterial for Efficient Removal of U(VI): Batch Experiments and Spectroscopy Study. *Sci. Bull.* **2018**, *63*, 831–839. [CrossRef]
51. Tomić, N.M.; Dohčević-Mitrović, Z.D.; Paunović, N.M.; Mijin, D.Ž.; Radić, N.D.; Grbić, B.V.; Aškrabić, S.M.; Babić, B.M.; Bajuk-Bogdanović, D.V. Nanocrystalline $CeO_{2-\delta}$ as Effective Adsorbent of Azo Dyes. *Langmuir* **2014**, *30*, 11582–11590. [CrossRef]
52. Ghoohestani, E.; Samari, F.; Homaei, A.; Yosuefinejad, S. A Facile Strategy for Preparation of Fe_3O_4 Magnetic Nanoparticles Using Cordia Myxa Leaf Extract and Investigating Its Adsorption Activity in Dye Removal. *Sci. Rep.* **2024**, *14*, 84. [CrossRef]
53. Di, J.; Ruan, Z.; Zhang, S.; Dong, Y.; Fu, S.; Li, H.; Jiang, G. Adsorption Behaviors and Mechanisms of Cu2+, Zn2+ and Pb2+ by Magnetically Modified Lignite. *Sci. Rep.* **2022**, *12*, 1394. [CrossRef] [PubMed]
54. Albadarin, A.B.; Mangwandi, C.; Al-Muhtaseb, A.H.; Walker, G.M.; Allen, S.J.; Ahmad, M.N.M. Kinetic and Thermodynamics of Chromium Ions Adsorption onto Low-Cost Dolomite Adsorbent. *Chem. Eng. J.* **2012**, *179*, 193–202. [CrossRef]
55. Dahlan, I.; Wan Mazlan, W.H.; Mulkan, A.; Zwain, H.M.; Hassan, S.R.; Aziz, H.A.; Hasan, H.Y.A.; Zekker, I. Modeling of Batch Organic Dye Adsorption Using Modified Metal-Organic Framework-5. *Chem. Eng. Technol.* **2022**, *45*, 2080–2087. [CrossRef]
56. Mohammadi, A.A.; Moghanlo, S.; Kazemi, M.S.; Nazari, S.; Ghadiri, S.K.; Saleh, H.N.; Sillanpää, M. Comparative Removal of Hazardous Cationic Dyes by MOF-5 and Modified Graphene Oxide. *Sci. Rep.* **2022**, *12*, 15314. [CrossRef]
57. Iprahim, M.; Farag, R.; Tantawe, H. Removal of Dyes from Aqueous Solution Using Zinc Terephthalic Acid Metal-organic Frameworks (Zn-TPA-MOF) Based Adsorbent. *Egypt. J. Chem.* **2022**, *66*, 295–305. [CrossRef]
58. Kavitha, D.; Namasivayam, C. Experimental and Kinetic Studies on Methylene Blue Adsorption by Coir Pith Carbon. *Bioresour. Technol.* **2007**, *98*, 14–21. [CrossRef]
59. Al-Rashed, S.M.; Al-Gaid, A.A. Kinetic and Thermodynamic Studies on the Adsorption Behavior of Rhodamine B Dye on Duolite C-20 Resin. *J. Saudi Chem. Soc.* **2012**, *16*, 209–215. [CrossRef]
60. Liu, X.; Gong, W.; Luo, J.; Zou, C.; Yang, Y.; Yang, S. Selective Adsorption of Cationic Dyes from Aqueous Solution by Polyoxometalate-Based Metal–Organic Framework Composite. *Appl. Surf. Sci.* **2016**, *362*, 517–524. [CrossRef]
61. Jiao, C.; Wang, Y.; Li, M.; Wu, Q.; Wang, C.; Wang, Z. Synthesis of Magnetic Nanoporous Carbon from Metal-Organic Framework for the Fast Removal of Organic Dye from Aqueous Solution. *J. Magn. Magn. Mater.* **2016**, *407*, 24–30. [CrossRef]
62. Budavari, S. (Ed.) *The Merck Index: An Encyclopedia of Chemicals, Drugs, and Biologicals*, 11th ed.; Centennial ed.; Merck: Rahway, NJ, USA, 1989; ISBN 978-0-911910-28-5.
63. Tong, M.; Liu, D.; Yang, Q.; Devautour-Vinot, S.; Maurin, G.; Zhong, C. Influence of Framework Metal Ions on the Dye Capture Behavior of MIL-100 (Fe, Cr) MOF Type Solids. *J. Mater. Chem. A* **2013**, *1*, 8534. [CrossRef]
64. Abdi, J.; Vossoughi, M.; Mahmoodi, N.M.; Alemzadeh, I. Synthesis of Metal-Organic Framework Hybrid Nanocomposites Based on GO and CNT with High Adsorption Capacity for Dye Removal. *Chem. Eng. J.* **2017**, *326*, 1145–1158. [CrossRef]
65. Zhao, X.; Liu, S.; Tang, Z.; Niu, H.; Cai, Y.; Meng, W.; Wu, F.; Giesy, J.P. Synthesis of Magnetic Metal-Organic Framework (MOF) for Efficient Removal of Organic Dyes from Water. *Sci. Rep.* **2015**, *5*, 11849. [CrossRef] [PubMed]
66. Huang, L.; He, M.; Chen, B.; Hu, B. Magnetic Zr-MOFs Nanocomposites for Rapid Removal of Heavy Metal Ions and Dyes from Water. *Chemosphere* **2018**, *199*, 435–444. [CrossRef]
67. Wang, K.; Li, C.; Liang, Y.; Han, T.; Huang, H.; Yang, Q.; Liu, D.; Zhong, C. Rational Construction of Defects in a Metal–Organic Framework for Highly Efficient Adsorption and Separation of Dyes. *Chem. Eng. J.* **2016**, *289*, 486–493. [CrossRef]
68. Yao, S.; Xu, T.; Zhao, N.; Zhang, L.; Huo, Q.; Liu, Y. An Anionic Metal–Organic Framework with Ternary Building Units for Rapid and Selective Adsorption of Dyes. *Dalton Trans.* **2017**, *46*, 3332–3337. [CrossRef] [PubMed]
69. Ahmad, N.; Younus, H.A.; Chughtai, A.H.; Van Hecke, K.; Khattak, Z.A.K.; Gaoke, Z.; Danish, M.; Verpoort, F. Synthesis of 2D MOF Having Potential for Efficient Dye Adsorption and Catalytic Applications. *Catal. Sci. Technol.* **2018**, *8*, 4010–4017. [CrossRef]
70. Huo, S.-H.; Yan, X.-P. Metal–Organic Framework MIL-100(Fe) for the Adsorption of Malachite Green from Aqueous Solution. *J. Mater. Chem.* **2012**, *22*, 7449. [CrossRef]
71. Chand, S.; Elahi, S.M.; Pal, A.; Das, M.C. A New Set of Cd(II)-Coordination Polymers with Mixed Ligands of Dicarboxylate and Pyridyl Substituted Diaminotriazine: Selective Sorption towards CO_2 and Cationic Dyes. *Dalton Trans.* **2017**, *46*, 9901–9911. [CrossRef]
72. Zhao, S.; Chen, D.; Wei, F.; Chen, N.; Liang, Z.; Luo, Y. Removal of Congo Red Dye from Aqueous Solution with Nickel-Based Metal-Organic Framework/Graphene Oxide Composites Prepared by Ultrasonic Wave-Assisted Ball Milling. *Ultrason. Sonochem.* **2017**, *39*, 845–852. [CrossRef]
73. Hazrati, M.; Safari, M. Cadmium-based Metal–Organic Framework for Removal of Dye from Aqueous Solution. *Environ. Prog. Sustain. Energy* **2020**, *39*, e13411. [CrossRef]
74. Yang, S.J.; Choi, J.Y.; Chae, H.K.; Cho, J.H.; Nahm, K.S.; Park, C.R. Preparation and Enhanced Hydrostability and Hydrogen Storage Capacity of CNT@MOF-5 Hybrid Composite. *Chem. Mater.* **2009**, *21*, 1893–1897. [CrossRef]
75. Yousef, N.S.; Farouq, R.; Hazzaa, R. Adsorption Kinetics and Isotherms for the Removal of Nickel Ions from Aqueous Solutions by an Ion-Exchange Resin: Application of Two and Three Parameter Isotherm Models. *Desalination Water Treat.* **2016**, *57*, 21925–21938. [CrossRef]
76. Lin, K.; Pan, J.; Chen, Y.; Cheng, R.; Xu, X. Study the Adsorption of Phenol from Aqueous Solution on Hydroxyapatite Nanopowders. *J. Hazard. Mater.* **2009**, *161*, 231–240. [CrossRef]

77. El-Nemr, M.A.; Yılmaz, M.; Ragab, S.; Hassaan, M.A.; El Nemr, A. Isotherm and Kinetic Studies of Acid Yellow 11 Dye Adsorptio from Wastewater Using Pisum Sativum Peels Microporous Activated Carbon. *Sci. Rep.* **2023**, *13*, 4268. [CrossRef]
78. Liu, Y.; Liu, Y.-J. Biosorption Isotherms, Kinetics and Thermodynamics. *Sep. Purif. Technol.* **2008**, *61*, 229–242. [CrossRef]
79. Lima, E.C.; Hosseini-Bandegharaei, A.; Anastopoulos, I. Response to "Some Remarks on a Critical Review of the Estimatio of the Thermodynamic Parameters on Adsorption Equilibria. Wrong Use of Equilibrium Constant in the van't Hoff Equatio for Calculation of Thermodynamic Parameters of Adsorption. *J. Mol. Liq.* **2019**, 273, 425–434". *J. Mol. Liq.* **2019**, *280*, 298–30 [CrossRef]
80. Liu, Y. Is the Free Energy Change of Adsorption Correctly Calculated? *J. Chem. Eng. Data* **2009**, *54*, 1981–1985. [CrossRef]
81. Liu, Y.; Xu, H. Equilibrium, Thermodynamics and Mechanisms of Ni^{2+} Biosorption by Aerobic Granules. *Biochem. Eng. J.* **200** *35*, 174–182. [CrossRef]
82. Garcia-Delgado, R.A.; Cotoruelo-Minguez, L.M.; Rodriguez, J.J. Equilibrium Study of Single-Solute Adsorption of Anioni Surfactants with Polymeric XAD Resins. *Sep. Sci. Technol.* **1992**, *27*, 975–987. [CrossRef]
83. Das, C.; Singh, S.; Bhakta, S.; Mishra, P.; Biswas, G. Bio-Modified Magnetic Nanoparticles with Terminalia Arjuna Bark Extract fo the Removal of Methylene Blue and Lead (II) from Simulated Wastewater. *Chemosphere* **2022**, *291*, 132673. [CrossRef] [PubMed

molecules

MDPI

Article

Green Ultrasound-Assisted Synthesis of Rare-Earth-Based MOFs

Francesca Lo Presti, Anna L. Pellegrino, Nancy Consoli and Graziella Malandrino *

Dipartimento di Scienze Chimiche, Università di Catania, and INSTM UdR Catania, Viale A. Doria 6, I-95125 Catania, Italy; francesca.lopresti@phd.unict.it (F.L.P.); annalucia.pellegrino@unict.it (A.L.P.)
* Correspondence: gmalandrino@unict.it

Abstract: Rare-earth (RE)-based metal organic frameworks (MOFs) are quickly gaining popularity as flexible functional materials in a variety of technological fields. These MOFs are useful for more than just conventional uses like gas sensors and catalyst materials; in fact, they also show significant promise in emerging technologies including photovoltaics, optical, and biomedical applications. Using yttrium and europium as ionic host centres and dopants, respectively, and 1,3,5-benzenetricarboxylic acid (H_3-BTC) as an organic linker, we describe a simple and green approach for the fabrication of RE-MOFs. Specifically, Y-BTCs and Eu-doped Y-BTCs MOFs have been synthesised in a single step using an eco-friendly method that makes use of ultrasound technology. To establish a correlation between the morphological and structural properties and reaction conditions, a range of distinct reaction periods has been employed for the synthetic processes. Detailed analyses of the synthesised samples through powder X-ray diffraction (PXRD), field emission scanning electron microscopy (FE-SEM), and Fourier-transform infrared spectroscopy (FT-IR) have confirmed the phase formation. Furthermore, thermal analyses such as thermogravimetric analysis (TGA) have been employed to evaluate the thermal stability and structural modifications of the Y-BTC and Eu-doped Y-BTC samples. Finally, the luminescent properties of the synthesised samples doped with Eu^{3+} have been assessed, providing an evaluation of their characteristics. As a proof of concept, an Eu-doped Y-BTC sample has been applied for the sensing of nitrobenzene as a molecule test of nitro derivatives.

Keywords: trimesic acid; yttrium-based MOF; morphological control; thermal stability; ultrasound; X-ray diffraction

1. Introduction

The scientific community has recently devoted a lot of interest to the study of metal-organic frameworks (MOFs), a unique class of materials that exhibit a number of fascinating properties. The ability to modify these features by appropriately altering the MOFs' structure, which can be changed by altering the metal centres, the organic linkers, and the synthesis' operational parameters, including temperature, process length, etc., is the primary reason for this interest [1]. A subclass of MOFs called luminescent MOFs has lately attracted a lot of attention because of their prospective applications in areas including sensing [2,3], lighting technologies [4], selective separation [5], and biological fields [6–10].

The primary luminescent emission mechanisms discovered for these compounds can be categorised into groups like organic linker-based emission, such as ligand-to-metal charge transfer (LMCT) or metal-to-ligand charge transfer (MLCT), metal-based emission, and antenna effect [11]. Specifically, the type of metal utilised determines most of the luminescent characteristics of produced MOFs; these substances emit light at certain wavelengths via a mechanism of metal atomic relaxation. For instance, rare-earth-based MOFs (RE-MOFs) exhibit bright luminescence as framework ions; nevertheless, the strength of their luminescence can be boosted via the antenna effect [12,13]. To date, many RE-MOFs have been made using various organic linkers [14–17]. In particular, due to the presence of oxygen atoms in carboxyl groups, organic linkers such as 1,3,5-benzenetricarboxylic acid (H_3-BTC) or 1,4-benzenedicarboxylic acid (H_2-BDC) are able to coordinate metal ions to form highly stable three-dimensional networks [18,19]. Regarding the central metal, Y^{3+}-based compounds are commonly used as matrices for traditional luminescent ions, such as Eu^{3+}, Tb^{3+}, and Tm^{3+} [20–23].

In regard to applications, RE-MOFs are now being developed as the most promising luminescent sensing materials for the detection of explosives [24], antibiotics [25], hazardous compounds and ions [26–28], heavy metals [29,30], and more. They have also been developed as catalysts [31], in thermometry [32,33], and in biological applications [34–38]. Additionally, the combination of RE-MOFs with polymeric structures could produce brand-new composite materials that merge the exceptional properties of MOFs with the features of the polymer supports [39–42].

Most RE-MOF synthetic approaches employed so far depend on solvothermal or hydrothermal techniques, which require high pressure, potentially hazardous organic solvents, and extremely lengthy reaction times [43–46]. There has not been too much effort, up to now, to change to a greener synthetic solution approach. For instance, Medina-Velazquez et al. [47] have developed Eu-BTC MOF nanopowders of high purity that are insoluble in water at room temperature in a basic medium. Zheng et al. [48] proposed the successful production of a variety of nano MOFs by chelating benzene tricarboxylic acid with various lanthanide ions on a wide scale by a one-step precipitation process in solution phase under ambient conditions. Recently, we performed a comprehensive study in which various reaction times and temperatures were tested in order to link the morphological and structural characteristics to the synthetic conditions [49]. This study has allowed us to better understand the mechanisms that would enable morphological control. Thus, while the morphology of the Y-BTC systems changes at higher synthetic temperatures with smaller and less homogenous crystals, the reaction time appears to favour a control in the development with an enhancement in morphological uniformity, both in terms of shape and size [49].

In the attempt to speed up the reaction process, reduce the reaction time, and switch to a greener method, several approaches have been developed, such as ultrasound (US) [50,51], microwave [52], mechanochemical synthesis [53,54], and electrodeposition [55].

In particular, the ultrasound method may increase the reactivity of the reagents, accelerate the crystallisation process, and produce homogeneous crystal particles with precise sizes [56]. In contrast, the reaction temperature and time, the solvents, and the process conditions have a strict influence on the grain size of the particles formed using techniques like hydrothermal or solvothermal synthesis.

Very few studies in the literature propose this convenient and fast synthetic method for the production of rare-earth-based MOFs [57–59]. More space is given to MOFs based on transition metals, such as Cu-BTC or even Zn-BTC, which are often used as catalysts [60,61].

In this study, we propose a green, facile, and rapid synthetic strategy to produce Y-BTC and Eu-doped Y-BTC MOFs using ultrasound (US) as a promising synthetic method. Specifically, the Y-BTC and Eu-doped Y-BTC MOFs have been synthesised under ultrasound at room temperature with reaction durations of 15, 30, 60, and 90 min. For the various Y-BTC and Eu-doped Y-BTCs samples, structural, thermal, and morphological characterisations have been carried out. Powder X-ray diffraction (PXRD), Fourier-transform infrared spectroscopy (FT-IR), thermogravimetric analyses (TGA), and differential scanning calorimetry (DSC) have all been used to evaluate the materials. Field emission scanning electron microscopy (FE-SEM) has been used to examine the morphology of powders that were directly deposited on aluminium stubs and sputtered with gold. This investigation provides a correlation between structure/morphology and synthetic process parameters. For the Eu-doped Y-BTC, preliminary luminescence characterisations and sensing of nitro derivatives have been carried out.

2. Results

2.1. MOF Synthesis

Using commonly accessible chemicals, water and ethanol as solvents, and mild pressure and temperature conditions, Y-MOFs have been synthesised through a straightforward ultrasound-assisted method. In several fields of chemical research, US-assisted syntheses have received a lot of interest. The possibility of carrying out chemical reactions at ambient temperature and pressure is one of the numerous advantages of ultrasonic technology. This ability is primarily attributed to the process of acoustic cavitation, which is the generation, growth, and collapse of bubbles, resulting in extraordinarily high local pressures and temperatures [62]. This approach has proven to be a successful means for synthesising MOFs at low reaction times, low reaction temperatures, and atmospheric pressure [57–59]. The syntheses have been carried out in water/ethanol solutions using 1,3,5-benzenetricarboxylic acid (H_3-BTC) as the organic binder, commonly known as trimesic acid, and $Y(CH_3COO)_3{\bullet}xH_2O$ and $Eu(CH_3COO)_3{\bullet}xH_2O$ as precursors of Y^{3+} and Eu^{3+}, the host metal centre and ionic dopant, respectively. Thus, two different types of systems have been synthesised: (i) pure Y-BTC MOFs with just yttrium as the metal centre, called Y-BTC; and (ii) Eu-doped Y-BTCs structures, named Y-BTC:Eu (10%). An Eu amount of 10% has been chosen to have enough luminescence signal for sensing applications, considering that usually pure Eu-MOFs are used for this purpose.

The syntheses were conducted under ultrasound at room temperature and atmospheric pressure for various reaction times according to the following reactions for Y-BTC (Equation (1)) and for Y-BTC:Eu (Equation (2)) and as reported in Scheme 1:

$$Y(CH_3COO)_3{\bullet}xH_2O + H_3 - BTC \overset{H_2O/EtOH}{\rightarrow} [Y(BTC){\cdot}6H_2O] + 3CH_3COOH \quad (1)$$

$$0.9\,Y(CH_3COO)_3{\bullet}xH_2O + 0.1\,Eu(CH_3COO)_3{\bullet}xH_2O + H_3 - BTC \overset{H_2O/EtOH}{\rightarrow} [Y_{0.9}Eu_{0.1}(BTC){\cdot}6H_2O] + 3CH_3COOH \quad (2)$$

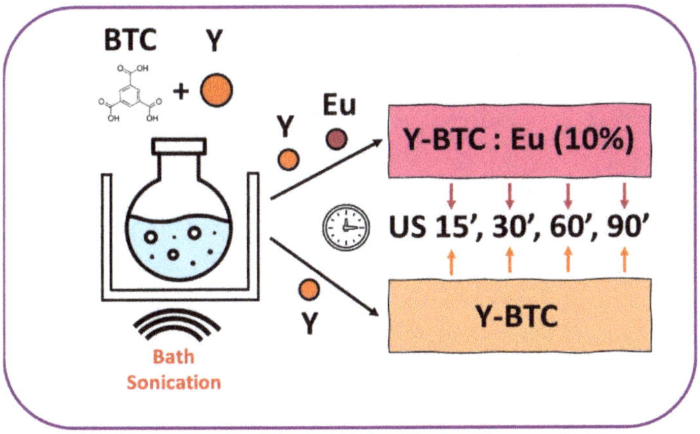

Scheme 1. Scheme of the various syntheses of Y-BTC and Y-BTC: Eu (10%) under ultrasound and at different reaction times (15–90 min).

2.2. *Structural and Morphological Characterisation*

For each synthesis of Y-MOFs and Y,Eu-MOFs, structural characterisations have been carried out concurrently through XRD and FE-SEM in order to ascertain if potential alterations or MOFs morphology changes are closely associated with their crystal structure. An overview of SEM images and XRD patterns for MOFs of Y-BTC:Eu (10%) made by ultrasonic-assisted synthesis is shown in Figure 1. The PXRD patterns of all Y-BTC:Eu (10%) syntheses are comparable. Specifically, for Y-BTC: Eu (10%) synthesised under ultrasound conditions for 15 (sample US 15′, reddish line) and 30 (sample US 30′, orange line) minutes, the crystal structures are consistent with the one described in ref. [63], which is coherent with the crystalline phase $Y(BTC)(H_2O)_6$. Figure S1 reports a comparison of the Y-BTC: Eu (10%) US 15′ with the $Y(BTC)(H_2O)_6$ (GOCYAY) structure [63]. This result finds a counterpart in the product observed in the green synthesis of Y-BTC MOFs at room temperature and atmospheric pressure in a water/ethanol solution [49]. Nonetheless, as the reaction time increases, peaks suggest the presence of a second phase that starts to develop alongside the main crystalline phase, e.g., see the Y-BTC:Eu (10%) US 60′ (blue line) and US 90′ (green line). This minority phase may be connected to a novel crystal structure recently described, different from all the other known structures, which consists of Y^{3+}/BTC^{3-} units and a very small amount of water, $[Y(BTC)]_3H_2O$. Such structure was produced by employing a synthesis under reflux that took place in a water/ethanol solution for 24 h at a temperature of 80 °C [49].

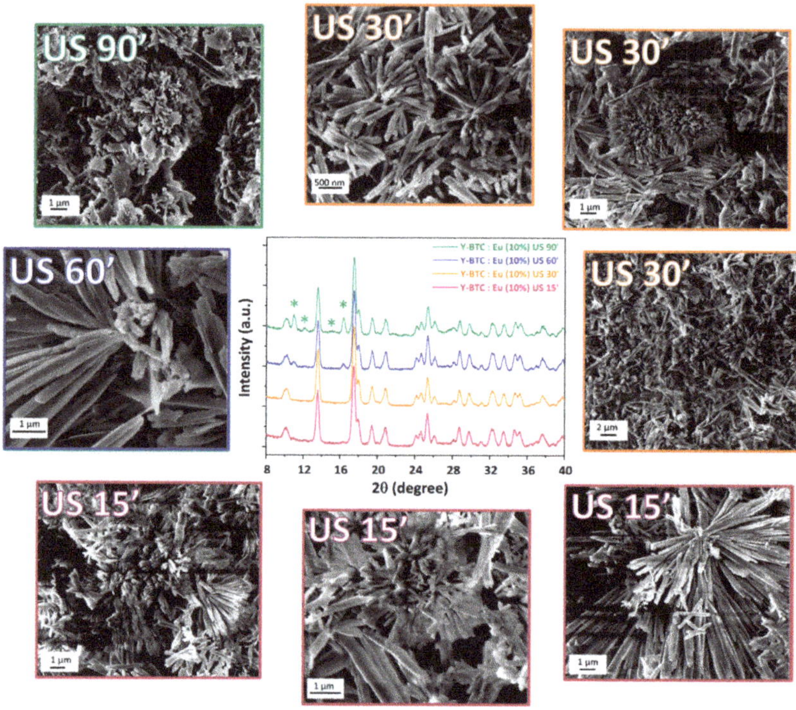

Figure 1. XRD patterns and FE-SEM images of Y-BTC:Eu (10%) samples produced under ultrasound with different reaction times: 15, 30, 60, and 90 min. * indicates peaks of the minority phase due to the novel crystal structure $[Y(BTC)]_3H_2O$ [49].

Time in ultrasonic synthesis also significantly affects morphology, paralleling structural changes. The FE-SEM images obtained for Y-BTC: Eu (10%) synthesised under US for 15 min (reddish line) indicate that the sample is extremely inhomogeneous, with rod-like structures in the order of several tens of microns, porous flower-like structures, needle-like crystals, and others in the shape of bundles of tubes. Thus, highly inhomogeneous structures of all sizes, from those that are nearly "macroscopic" to those with dimensions in the tens of nanometers and a high aspect ratio, are present as a result of the first formed product. A reduction in morphological and dimensional inhomogeneity may be seen by reaching 30 min (Y-BTC: Eu (10%) US 30′—orange line), in which the flower and stick shapes are maintained. A further improvement in morphological and dimensional homogeneity is shown in the sample synthesised for 60 min (Y-BTC: Eu (10%) US 60′—blue line), with elongated stick-shaped structures having lengths on the order of tens of microns. With these reaction times, smaller, flattened, sheet-like structures also start to emerge. After 90 min of ultrasound, the morphology finally seems more altered than one may expect: the huge sticks start to break, and the leaflet morphologies become more apparent and bigger (US 90′, green line).

This alteration in morphology is likely to be connected to the structural alterations of Y-BTC:Eu (10%); at high synthetic times (US 60′ and US 90′), the variation in the structure and, therefore, in the morphology, suggests that it may be linked to the synthetic process. The ultrasonic cavity procedure might produce free radicals that cause the material's structure to change as a result of the high pressure and high temperature conditions that are locally present in the reaction environment. The frequency of ultrasound-induced collisions increases with time, accelerating the interaction between the substrate and the reactant and likely leading to structural and morphological changes.

Energy dispersive X-ray analysis has also been performed on all the Eu-doped Y-BTC samples to gather qualitative and quantitative data on the presence of europium in the structures. As a proof case, the EDX spectrum of the US 90′ sample is reported in Figure S2. The X-ray L lines of Eu and Y are observed in the range 5.8–6.5 keV and at 1.92 keV, respectively, in addition to the C K_α peak found at 0.28 keV and the O K_α peak at 0.524 keV. In particular, the quantitative analysis gives a percentage of Eu doping in the sample of 9.5 ± 0.5%.

A similar situation in terms of structural and morphological characteristics is found for the undoped Y-BTC MOF synthesised by ultrasound for 15, 30, 60, and 90 min. The XRD patterns of Y-BTC samples synthesised at various times are displayed in Figure 2. For the undoped samples, it is also evident that the Y-BTC US 15′ sample produced a pattern analogous to the structure of ref. [63]. The two samples of Y-BTC and Y-BTC: Eu (10%) synthesised for 15′ appear to be isostructural with the GOCYAY structure [63], in which yttrium is nona-coordinated by six oxygen atoms from water molecules as well as three oxygen atoms from three carboxylate groups to form a tricapped trigonal prism (Figure S3). A comparison of the three patterns of Y-BTC US 15′, Y-BTC: Eu US 15′, and GOCYAY is reported in Figure S1. On increasing reaction time, peaks corresponding to a second crystalline phase appear for samples Y-BTC US 30′, 60′, and 90′, analogously to what is observed for the Eu-doped Y-BTC MOFs.

The morphology of the US samples of Y-BTC similarly displays the same characteristics as those of the doped samples. In fact, the Y-BTC US 15′ MOF (pink line) mostly depicts heterogeneous structures, with a tendency for stick-like growths of various sizes and shapes to eventually take on the appearance of flowers. The morphological inhomogeneity starts to reduce as the reaction time increases (US 30′—blue line), and the sticks that have developed have thicker and longer lengths. Furthermore, sheet-like forms start to emerge. The MOF sticks keep expanding in size and form for the Y-BTC US 60′ sample (green line), reaching lengths in the tens of microns. The stick structures finally seemed fractured after reaction times of 90 min (blue line), so the sticks changed into bigger sheet-like shapes.

The FT-IR spectra of the Y-BTC and Y-BTC: Eu (10%) samples synthesised for 15 and 90 min provide additional support to the observations made by PXRD patterns. The FT-IR spectra of Y-BTC US 15′ and Y-BTC US 90′ are shown in Figure 3a. According to the degree and type of hydration in the two systems, these spectra demonstrate the existence of multiple peaks in the 3200–3500 cm^{-1} region, more pronounced for the US 15′ sample, associated with the -OH stretching. The nujol C–H stretching causes the signal at 2900 cm^{-1} to appear. Peaks at 1617 and 1560 cm^{-1} are related to the vibrations of the BTC ligand, and in particular to the stretching of the C=O and the asymmetric stretching of the COO^-, respectively. The symmetric stretching of the COO^-, usually observed around 1400 cm^{-1}, is buried under the peaks due to the nujol oil at 1458 and 1375 cm^{-1}.

Additional peaks seen in the fingerprint region (1000–700 cm^{-1}) can be accounted for by the vibrational modes of the benzene rings and, in particular, the C-H bending vibrations. As the PXRD patterns revealed a similar crystalline phase, it is expected that the FT-IR spectra of the Y-BTC:Eu (10%) samples are equivalent to those of the parent undoped ones. Figure 3b shows the FT-IR spectra of the Eu-doped Y-BTCs synthesised for 15 and 90 min. A sequence of peaks at 3323, 3406, and 3485 cm^{-1}, are present in the area of -OH stretching, suggesting the presence of water molecules coordinated to the metal ion (either Y or Eu). Peaks at 1614 and

1562 cm^{-1} represent the C=O stretching vibration and the COO$^-$ asymmetric stretching, respectively, associated with the BTC ligand inside the MOF structure. Peaks related to the C–H bending vibrations are seen in the range of 1000–700 cm^{-1}.

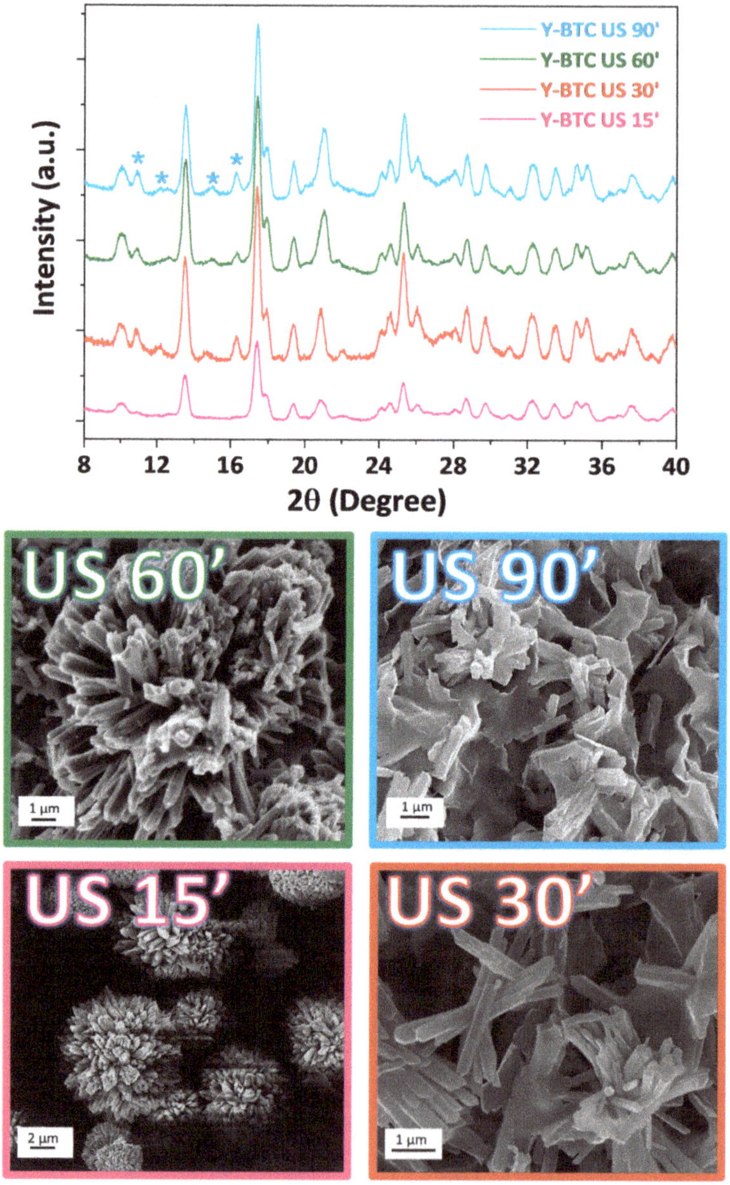

Figure 2. XRD patterns and FE-SEM images of Y-BTC samples synthetised under ultrasound for different reaction times: 15, 30, 60, and 90 min. * indicates peaks of a minority phase due to the novel crystal structure [Y(BTC)]$_3$H$_2$O [49].

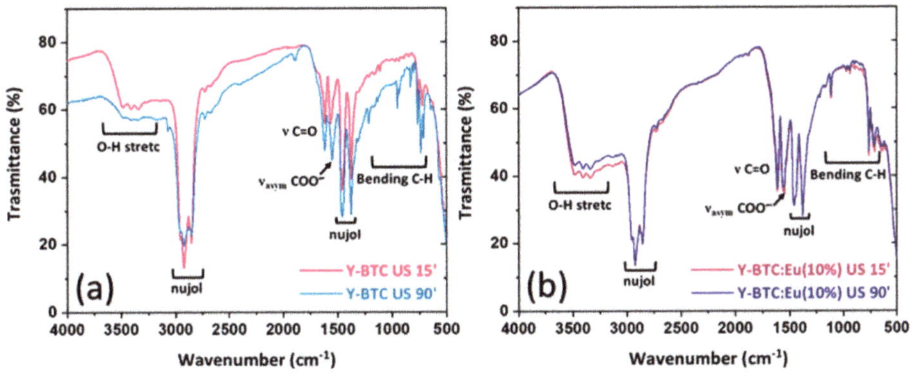

Figure 3. FT-IR spectra of (**a**) Y-BTC US 15′ and Y-BTC US 90′; (**b**) Y-BTC: Eu (10%) US 15′ and Y-BTC: Eu (10%) US 90′.

2.3. Thermal Properties

Thermogravimetric measurements have been used to investigate the Y-BTC MOF's thermal stability and confirm the presence of water molecules. As a result, the thermal characterisation of these Y-BTC samples is useful in identifying the potential presence and quantity of coordinated water molecules inside the MOF structure. The TGA curves of the Y-BTC US 15′ (pink line) and Y-BTC US 90′ are shown in Figure 4a (light blue line). For the Y-BTC US 15′, a first weight loss step of about 6.9% can be seen at 66 °C, which is probably the result of the loss of the first water molecule per unit formula $Y(BTC)(H_2O)_6$, while a second, larger weight loss step of 26.1% can be seen at 125 °C, which is the result of the loss of the additional five H_2O molecules that coordinate the metal centre. The Y-BTC US 90′ exhibits a similar situation, showing three distinct weight loss steps of 4.4%, 8.4%, and 23.1% at temperatures of 62 °C, 86 °C, and 126 °C, respectively. These steps correspond to the loss of roughly 5.4 H_2O molecules over one Y(BTC) unit.

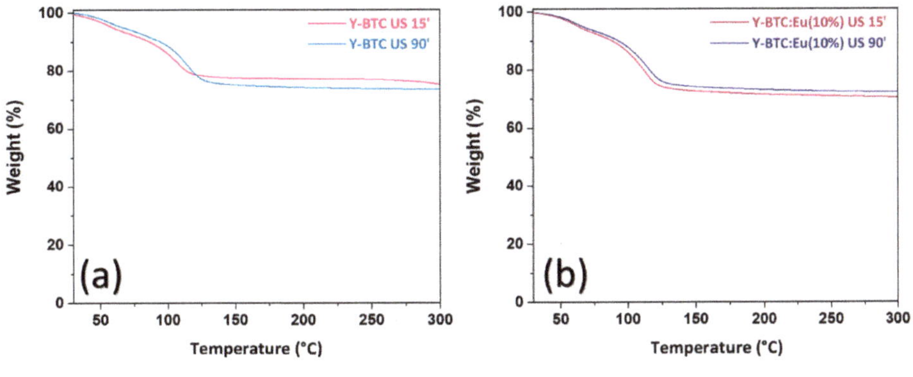

Figure 4. TGA curves of (**a**) Y-BTC US 15′ and Y-BTC US 90′; (**b**) Y-BTC: Eu (10%) US 15′ and Y-BTC: Eu (10%) US 90′.

The curves show similar thermal behaviour for the Y-BTC: Eu samples (Figure 4b). In particular, the graphs reveal two distinct weight losses at 72 °C and 120 °C. The first step often results in the loss of 1.5 molecules of water per unit formula, whereas the second step removes all 4 molecules of water that are still present, leaving behind a residue that is compatible with anhydrous Y-BTC.

The results of the thermal analysis are consistent with the infrared spectra, which show peaks due to numerous -OH stretching, and with the XRD patterns, which show for all the pure and Eu-doped Y-BTC samples a pattern corresponding to the $Y(BTC)(H_2O)_6$ structure.

2.4. Luminescence Characterisation

Using dispersions of the Y-BTC: Eu (10%), US 15′, US 30′, US 60′, and US 90′ samples in aqueous media, photoluminescence measurements have been used to perform preliminary luminescent characterisation (see Figure 5). A slight effect of the MOF's crystal structure is visible in the luminescence intensity of the various samples. Thus, no difference is found for the samples produced at 15, 30, and 60 min, where just one, or at least a dominant, crystalline phase is visible. In any case, the presence of a second crystalline phase in the sample produced at 90 min does not significantly affect the luminescence behaviour of the sample. At an excitation wavelength of 265 nm, all Eu-doped Y-MOF exhibit fluorescence emission. In Figure 5, the spectra of the Y-BTC: Eu (10%) US 15′ and Y-BTC: Eu (10%) US 90′ are reported. The strong peaks centred at 592 and 618 nm may be attributed to the $^5D_0 \rightarrow {}^7F_1$ and $^5D_0 \rightarrow {}^7F_2$ transitions, respectively, which represent two of the typical emissions of the Eu^{3+} ion. It is interesting to evaluate the asymmetry ratio *R*, i.e., the ratio between the integrated areas of the $^5D_0 \rightarrow {}^7F_2$ and $^5D_0 \rightarrow {}^7F_1$ electronic transitions, which can give a hint on the symmetry of the Eu^{3+} ion. In both cases, the R values of 1.43 and 1.65 for the Y-BTC: Eu (10%) US 15′ and Y-BTC: Eu (10%) US 90′, respectively, are indicative of a quite symmetric environment [64] for the Eu^{3+} ion, as expected for a tri-capped trigonal prismatic coordination, the coordination of europium when substituting yttrium in the $Y(BTC)(H_2O)_6$ structure (Figure S3).

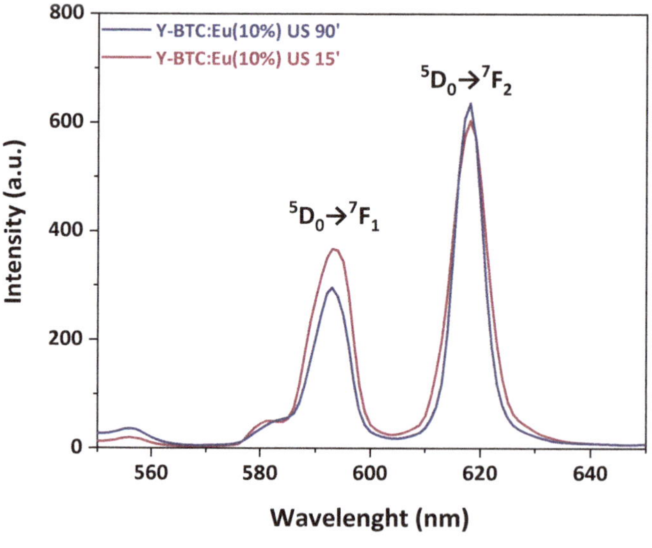

Figure 5. Luminescence spectra of Y-BTC US 15′ and Y-BTC US 90′, $\lambda_{exc.}$ = 265 nm.

The Y-BTC: Eu (10%) sample synthesised under ultrasound may hold promise as a potential luminescent material for different applications and other practical uses due to its strong emission, even for samples produced with very short reaction times.

2.5. Sensing of Nitro Derivatives

The Y-BTC: Eu (10%) samples synthesised under ultrasound may hold promise as potential systems for the detection of nitro derivatives.

Nitroaromatic chemicals are the principal components of explosives used in acts of terrorism and in improvised explosive devices. Thus, there is a critical need to develop reliable, user-friendly sensing materials for identifying nitroaromatic pollutants. As a proof of concept, the Y-BTC: Eu (10%) MOF 90′ sample, chosen because it is well crystallised and has a quite homogeneous morphology, has been applied as a fluorescent probe for sensing nitrobenzene ($C_6H_5NO_2$ or NB), a test molecule for nitro derivatives.

Aqueous solutions have been used to disperse the as-prepared Y-BTC:Eu MOF US 90′ (1 mg in 2 mL of H_2O), and different quantities of $C_6H_5NO_2$ (10^{-6} M solution in H_2O) have been added to the suspension. To guarantee equal distribution and avoid sedimentation effects, the suspension has been sonicated for five minutes after each addition to produce suspensions suitable for luminescent measurements. The choice of a 5 min period of sonication allowed for the best possible interaction between the MOF and the identified species. Figure 6a illustrates a pronounced quenching effect on both the $^5D_0 \rightarrow {}^7F_1$ and $^5D_0 \rightarrow {}^7F_2$ peak intensities of the Eu^{3+} emission, which becomes evident even after the addition of just 10 μL of NB contaminant. To quantitatively assess this phenomenon, the Stern–Volmer (SV) equation $I_0/I = 1 + K_{sv}[M]$ [65] has been employed. The I_0 and I represent the intensity values of the most intense peaks (617.5 nm) of the Y-BTC: Eu spectrum in the absence and presence of analytes, respectively. K_{sv} is the quenching constant, and [M] denotes the absolute concentration of the analyte. The linear coefficients R, close to one, reveal strong linear correlations and a good fit for the Stern–Volmer equation model.

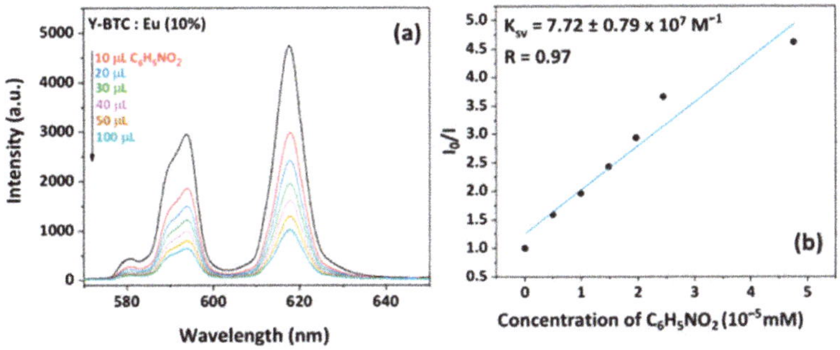

Figure 6. (**a**) Emission spectra and (**b**) K_{sv} curve of Y-BTC: Eu (10%) dispersed into aqueous solution (1 mg in 2 mL) in the presence of various concentrations of NB under excitation at 265 nm.

Furthermore, using the K_{sv} equation, the quenching impact of the aromatic component nitrobenzene is expected to be $7.72 \pm 0.79 \times 10^7$ M^{-1}, indicating a very strong quenching effect on the Eu^{3+} luminescence. The detection mechanism behind this intriguing behaviour is attributed to the photo-induced intermolecular electron transfer between the MOF sensor and the analyte, as assessed through density functional theory (DFT) and time-dependent DFT [66,67].

The presently found K_{sv} value is higher than most of the previous works [68–72], proving the excellent sensitivity of the Eu-doped Y-BTC MOF to detect nitroaromatic compounds.

These findings demonstrate a promising outcome considering that presently only a 10% Eu-doped sample has been used, making the testing technique less expensive than using pure Eu-MOF structures.

3. Discussion

The ultrasound-assisted synthetic approach described in this study represents a green, quick, facile, and reproducible preparation method that can be used for the synthesis of Y-based MOFs in contrast to the existing synthetic methods reported in the literature, which frequently involve numerous steps, high temperatures, toxic solvents, and long reaction times. The US synthetic method has been demonstrated to be an effective approach for producing MOFs with fast crystallisation times, low reaction temperatures, and atmospheric pressure. This straightforward but effective synthetic method may be compared with that described in ref. [59], which describes the production of isostructural Ln(BTC) MOFs

(Ln = Ce, Tb, and Y) with a tetragonal structure at room temperature through ultrasonic irradiation. In contrast to the previous study that focused on the minimum times for the full crystallisation of the Ln-MOF, the reaction conditions used in the present study have made it possible to conduct a more in-depth analysis of the crystallisation process and correlate the morphology and structure of the various samples to reaction conditions. The dopant's contribution to both structural and luminescence properties of the Y-MOF structure could also be examined. For instance, it was shown that regardless of the dopant's presence, the sample crystallinity tends to increase as the reaction time increases. The Y-MOF crystal structure of all the synthesised samples with different reaction times is comparable to that of ref. [63]. However, as reaction time increases, peaks that correspond to a second crystalline form start to develop. Such a new secondary phase consists of Y^{3+}/BTC^{3-} units and very little water. From a morphological perspective, reaction time in ultrasonic synthesis promotes an improvement in shape distribution and a decrease in length that is more evident at 90 min of reaction. In fact, several types of morphologies, such as rod-like, flower-like, and needle-like structures, may be observed for Y-MOFs, both doped and undoped, prepared at short reaction durations, i.e., 15 and 30 min. The 60 min samples have a more uniform morphology, with stick-shaped structures on the order of tens of microns. The luminescent features of these MOFs, caused by the presence of Eu ions introduced as dopants into the Y-BTC structures, partly correlate with the synthesis time and, therefore, the crystallinity of the final samples. The luminescence behaviour of the Y-BTC: Eu (10%) US 15′ and 90′ is similar, indicating that the fluctuation in the MOF crystal structure does not significantly affect the emission of the Eu ion.

4. Materials and Methods

4.1. Materials

Yttrium(III) acetate hydrate and Europium(III) acetate hydrate were purchased from STREM Chemicals, and the 1,3,5-benzenetricarboxylic acid was purchased from Sigma-Aldrich. All the reagents were used without other purifications. Two different sets of syntheses have been performed. Scheme 1 provides a summary of the process.

4.2. Synthesis under Ultrasound

Y-BTC: Eu (10%). 50 mL of a water solution containing 3.6 mmol of $Y(CH_3COO)_3{\bullet}xH_2O$ and 0.4 mmol of $Eu(CH_3COO)_3{\bullet}xH_2O$ were added to a glass balloon with a grounded neck. After that, the solution was mixed with 50 mL of ethanol containing H_3-BTC (4 mmol), and the system was exposed to ultrasound for various lengths of time: 15, 30, 60, and 90 min. The products were simply filtered out after the various reaction periods, and the Eu-doped Y-BTC precipitate was recovered, cleaned with water/ethanol, and allowed to dry in the open air.

Y-BTC. The undoped chemical products were obtained using a procedure similar to that used to create the related Eu-doped Y-BTC product, starting with 50 mL of a water solution containing 4 mmol of $Y(CH_3COO)_3{\bullet}xH_2O$ and 50 mL of ethanol containing 4 mmol of H_3-BTC acid. The durations of the ultrasonic treatments on the system were 15, 30, 60, and 90 min, respectively. The product was filtered before being rinsed with water and ethanol and dried in the air.

4.3. Characterisations

Fourier transform infrared (FT-IR) spectra on a JASCO FTIR 4600 LE spectrometer (Easton, MD, USA) were collected. A small amount of sample powder had been thoroughly pulverised in an agate mortar in nujol and placed between NaCl plates, which underwent FT-IR analysis. The Mettler Toledo TGA2 and STARe software were used to conduct thermogravimetric experiments. A 50 sccm pure nitrogen flow, atmospheric pressure, and a 5 °C/min heating rate were used in the dynamic thermal investigations. The weights of the samples were around 10 mg. With a Smartlab (Rigaku, Tokyo, Japan) diffractometer, XRD patterns were captured using a rotating Cu K_{α} anode operating at 45 kV and 200 mA. During the acquisition, a 0.02° increment step was used. The

ZEISS SUPRA 55VP field-emission scanning electron microscope was used to carry out the morphological characterisation. Samples were attached to Al stubs using graphite double-sided adhesives before being sputtered with Au in order to characterise the morphology of the synthesised Y-MOF using the FE-SEM apparatus. Using energy dispersive X-ray (EDX) analysis by means of an INCA-Oxford windowless detector with a resolution of 127 eV, which was calculated as the full width half maximum (FWHM) of the Mn K_α, the atomic composition of the films was determined. A JASCO FP-8300 spectrofluorimeter was used to capture photoluminescence spectra at room temperature with an excitation wavelength of 265 nm. In 2 mL of water, 2 mg of Y-BTC: Eu (10%) was dispersed for each experiment. The sensing test was carried out by adding different amounts of $C_6H_5NO_2$ (10^{-6} M in H_2O) to the Y-BTC: Eu (10%) dispersion (1 mg in 2 mL of H_2O). Before starting additions, the Y-BTC: Eu dispersion was sonicated for half an hour to have a stable suspension. After each addition, the resulting mixtures were sonicated for 5 min to favour interaction and prepare them for luminescence measurements. The sequential $C_6H_5NO_2$ detection tests used the same Y-BTC: Eu MOF powder.

5. Conclusions

In summary, this research study offers a simple, quick, and environmentally friendly method for creating several luminescent Eu-doped Y-BTC and undoped Y-BTC MOFs. Notably, the syntheses have applied ultrasonic technology under standard pressure and room temperature while just changing the reaction time. The extensive characterisations conducted in this work consistently demonstrate a repeatable process where the structural properties of the Y-BTCs and the synthetic process parameters are closely related. From a structural point of view, each sample synthesised using different reaction times displays a dominant phase linked to the crystalline $Y(BTC)(H_2O)_6$ MOF. However, as the reaction time increases, a secondary phase that shows the Y^{3+}/BTC^{3-} units and a little quantity of water in the crystalline phase appears. Numerous morphologies, including porous flower-like formations, needle-shaped crystals, and tube-like structures, were seen in those synthesised with shorter reaction times (i.e., 15 and 30 min). However, as the reaction time increased, some of these structures began to sinter, giving rise to filamentous or tubular structures with higher aspect ratios. Eventually, the structures completely broke down after very lengthy reaction periods, like 90 min. Additionally, the 15 and 90 min samples have been shown to be highly stable, after losing the coordinated water molecules, up to 400 °C. The Eu-doped Y-BTC samples have comparable luminescent properties. This study represents a facile, general route to synthesise a variety of RE-doped MOFs, envisaging the possibility of tuning the specific properties in function of the desired sensing applications.

Supplementary Materials: The following supporting information can be downloaded at: https://www.mdpi.com/article/10.3390/molecules28166088/s1, Figure S1: Comparison of the Y-BTC US 15′, Y-BTC: Eu US 15′ and GOCYAY-$Y(BTC)(H_2O)_6$ structures; Figure S2: EDX spectrum of the Y-BTC: Eu US 90′ sample; Figure S3: Coordination of the Y in the $Y(BTC)(H2O)_6$ structure.

Author Contributions: Methodology, F.L.P.; data curation, F.L.P. and A.L.P.; validation, A.L.P.; investigation, F.L.P. and N.C.; validation, G.M.; writing—original draft preparation, F.L.P.; writing—review and editing, G.M.; funding acquisition, G.M. All authors have read and agreed to the published version of the manuscript.

Funding: This research was funded by the University of Catania within the PIACERI research program UNICT 2020-22 Linea 2.

Institutional Review Board Statement: Not applicable.

Informed Consent Statement: Not applicable.

Data Availability Statement: Data are contained within the article or Supplementary Materials.

Acknowledgments: The authors thank the Bionanotech Research and Innovation Tower (BRIT) laboratory of the University of Catania (Grant no. PONa3_00136 financed by the Italian Ministry for

Education, University, and Research, MIUR) for the diffractometer facility. A. L. P. thanks the Ministero dell'Università e della Ricerca within the PON "Ricerca e Innovazione" 2014-2020 Azioni IV.4 program.

Conflicts of Interest: The authors declare no conflict of interest.

References

1. Wang, Q.; Astruc, D. State of the Art and Prospects in Metal–Organic Framework (MOF)-Based and MOF-Derived Nanocatalysis. *Chem. Rev.* **2020**, *120*, 1438–1511. [CrossRef]
2. Hao, Z.; Song, X.; Zhu, M.; Meng, X.; Zhao, S.; Su, S.; Yang, W.; Songa, S.; Zhang, H. One-dimensional channel-structured Eu-MOF for sensing small organic molecules and Cu^{2+} ion. *J. Mater. Chem. A* **2013**, *1*, 11043–11050. [CrossRef]
3. Barroso, N.; Andreo, J.; Beobide, G.; Castillo, O.; Luque, A.; Pérez-Yáñez, S.; Wuttke, S. Magnetic sustentation as an adsorption characterization technique for paramagnetic metal-organic frameworks. *Commun. Chem.* **2023**, *6*, 4. [CrossRef] [PubMed]
4. Karbalaee Hosseini, A.; Tadjarodi, A. Luminescent Cd coordination polymer based on thiazole as a dual-responsive chemosensor for 4-nitroaniline and CrO_4^{2-} in water. *Sci. Rep.* **2023**, *13*, 269. [CrossRef]
5. Ma, D.; Li, Z.; Zhu, J.; Zhou, Y.; Chen, L.; Mai, X.; Liufu, M.; Wu, Y.; Li, Y. Inverse and highly selective separation of CO_2/C_2H_2 on a thulium–organic framework. *J. Mater. Chem. A* **2020**, *8*, 11933–11937. [CrossRef]
6. Stillman, Z.S.; Decker, G.E.; Dworzak, M.R.; Bloch, E.D.; Fromen, C.A. Aluminum-based metal–organic framework nanoparticles as pulmonary vaccine adjuvants. *J. Nanobiotechnol.* **2023**, *21*, 39. [CrossRef] [PubMed]
7. Chang, Y.; Chen, H.; Xie, X.; Wan, Y.; Li, Q.; Wu, F.; Yang, R.; Wang, W.; Kong, X. Bright Tm^{3+}-based downshifting luminescence nanoprobe operating around 1800 nm for NIR-IIb and c bioimaging. *Nat. Commun.* **2023**, *14*, 1079. [CrossRef]
8. Zhong, Y.; Peng, Z.; Peng, Y.; Li, B.; Pan, Y.; Ouyang, Q.; Sakiyama, H.; Muddassir, M.; Liu, J. Construction of Fe-doped ZIF-8/DOX nanocomposites for ferroptosis strategy in the treatment of breast cancer. *J. Mater. Chem. B* **2023**, *11*, 6335–6345. [CrossRef]
9. Chen, X.; Li, M.; Lin, M.; Lu, C.; Kumar, A.; Pan, Y.; Liu, J.; Peng, Y. Current and promising applications of Hf(IV)-based MOFs in clinical cancer therapy. *J. Mater. Chem. B* **2023**, *11*, 5693–5714. [CrossRef]
10. Xu, Z.; Wu, Z.; Huang, S.; Ye, K.; Jiang, Y.; Liu, J.; Liu, J.; Lu, X.; Li, B. A metal-organic framework-based immunomodulatory nanoplatform for anti-atherosclerosis treatment. *J. Control. Release* **2023**, *354*, 615–625. [CrossRef]
11. Allendorf, M.D.; Bauer, C.A.; Bhaktaa, R.K.; Houka, R.J.T. Luminescent metal–organic frameworks. *Chem. Soc. Rev.* **2009**, *38*, 1330–1352. [CrossRef] [PubMed]
12. Soares-Santos, P.C.R.; Cunha-Silva, L.; Paz, F.A.A.; Ferreira, R.A.S.; Rocha, J.; Trindade, T.; Carlos, L.D.; Nogueira, H.I.S. Hydro-Ionothermal Synthesis of Lanthanide-Organic Frameworks with 1,4-Phenylenebis(methylene)diphosphonate. *Cryst. Growth Des.* **2008**, *8*, 2505–2516. [CrossRef]
13. Hu, S.; Liu, J.; Wang, Y.; Liang, Z.; Hu, B.; Xie, J.; Wong, W.-L.; Wong, K.-Y.; Qiu, B.; Peng, W. A new fluorescent biosensor based on inner filter effect and competitive coordination with the europium ion of non-luminescent Eu-MOF nanosheets for the determination of alkaline phosphatase activity in human serum. *Sens. Actuators B Chem.* **2023**, *380*, 133379. [CrossRef]
14. Hooda, A.; Singh, D.; Dalal, A.; Nehra, K.; Kumar, S.; Singh Malik, R.; Kumarc, R.; Kumarc, P. Optical, electrochemical and photophysical analyses of heteroleptic luminescent Ln(III) complexes for lighting applications. *RSC Adv.* **2023**, *13*, 9033–9045. [CrossRef]
15. Main, R.M.; Vornholt, S.M.; Rice, C.M.; Elliott, C.; Russell, S.E.; Kerr, P.J.; Warren, M.R.; Morris, R.E. In situ single-crystal synchrotron X-ray diffraction studies of biologically active gases in metal-organic frameworks. *Commun. Chem.* **2023**, *6*, 44. [CrossRef]
16. Liu, M.; Shang, C.; Zhao, T.; Yu, H.; Kou, Y.; Lv, Z.; Hou, M.; Zhang, F.; Li, Q.; Zhao, D.; et al. Site-specific anisotropic assembly of amorphous mesoporous subunits on crystalline metal–organic framework. *Nat. Commun.* **2023**, *14*, 1211. [CrossRef]
17. Shah, A.H.; Ul Abideen, Z.; Maqsood, S.; Rashid, F.; Ullah, R.; Ur Rehman, A.; Dildar, M.; Ahmad, M.; Ullah, K.; Rafi, M.N.; et al. Porous Cu-based metal organic framework (Cu-MOF) for highly selective adsorption of organic pollutants. *J. Solid State Chem.* **2023**, *322*, 123935. [CrossRef]
18. Feng, J.; Yang, X.; Gao, S.; Shi, J.; Cao, R. Facile and Rapid Growth of Nanostructured Ln-BTC Metal–Organic Framework Films by Electrophoretic Deposition for Explosives sensing in Gas and Cr^{3+} Detection in Solution. *Langmuir* **2017**, *33*, 14238–14243. [CrossRef]
19. Yang, X.P.; Jones, R.A.; Riversa, J.H.; Lai, R.P. Syntheses, structures and luminescent properties of new lanthanide-based coordination polymers based on 1,4-benzenedicarboxylate (bdc). *Dalton Trans.* **2007**, *15*, 3936–3942. [CrossRef]
20. Zhang, Y.; Li, B.; Ma, H.; Zhang, L.; Zhang, W. An RGH–MOF as a naked eye colorimetric fluorescent sensor for picric acid recognition. *J. Mater. Chem. C* **2017**, *5*, 4661–4669. [CrossRef]
21. Alammar, T.; Hlova, I.Z.; Gupta, S.; Balema, V.; Pecharskya, V.K.; Mudring, A.V. Luminescence properties of mechanochemically synthesized lanthanide containing MIL-78 MOFs. *Dalton Trans.* **2018**, *47*, 7594–7601. [CrossRef] [PubMed]
22. Almáši, M.; Zeleňák, V.; Kuchár, J.; Bourrelly, S.; Llewellyn, P.L. New members of MOF-76 family containing Ho(III) and Tm(III) ions: Characterization, stability and gas adsorption properties. *Colloids Surf.* **2016**, *496*, 114–124. [CrossRef]
23. Yang, X.; Lin, X.; Zhao, Y.; Zhao, Y.S.; Yan, D. Lanthanide Metal–Organic Framework Microrods: Colored Optical Waveguides and Chiral Polarized Emission. *Angew. Chem. Int. Ed.* **2017**, *129*, 7961–7965. [CrossRef]

24. Zhao, Z.; Liu, L. Multiple fluorescent sensing responses of a 2D framework structure towards explosive: Synthesis, characterization and sensing performance. *J. Mol. Struct.* **2021**, *1238*, 130418. [CrossRef]
25. Yang, Y.; Zhao, L.; Sun, M.; Wei, P.; Li, G.; Li, Y. Highly sensitive luminescent detection toward polytypic antibiotics by a water-stable and white-light-emitting MOF-76 derivative. *Dye. Pigment.* **2020**, *180*, 108444. [CrossRef]
26. Dong, X.; He, Q.; Li, M.; Wang, X.; Wang, Y.; Zhang, W. Fluorescence and electrochemical detection of iodine vapor in the presence of high humidity using Ln-based MOFs. *Dalton Trans.* **2021**, *50*, 15567–15575. [CrossRef]
27. Wu, R.Z.; Yang, X.; Zhang, L.W.; Zhou, P.P. Luminescent lanthanide metal–organic frameworks for chemical sensing and toxic anion detection. *Dalton Trans.* **2017**, *46*, 9859–9867. [CrossRef]
28. Puglisi, R.; Pellegrino, A.L.; Fiorenza, R.; Scirè, S.; Malandrino, G. A Facile One-Pot Approach to the Synthesis of Gd-Eu Based Metal-Organic Frameworks and Applications to Sensing of Fe^{3+} and $Cr_2O_7^{2-}$ Ions. *Sensors* **2021**, *21*, 1679. [CrossRef]
29. Jamali, A.; Tehrani, A.A.; Shemirani, F.; Morsali, A. Lanthanide metal–organic frameworks as selective microporous materials for adsorption of heavy metal ions. *Dalton Trans.* **2016**, *45*, 9193–9200. [CrossRef]
30. Lian, X.; Yan, B. A lanthanide metal–organic framework (MOF-76) for adsorbing dyes and fluorescence detecting aromatic pollutants. *RSC Adv.* **2016**, *6*, 11570–11576. [CrossRef]
31. Wang, L.; He, J.; Chen, X.; Lv, Y. A lanthanide MOF catalyst with an excellent thermal stability for the synthesis of polycarbonate diol. *J. Iran. Chem. Soc.* **2020**, *17*, 2335–2343. [CrossRef]
32. Liu, J.; Zhao, Y.; Li, X.; Wu, J.; Han, Y.; Zhang, X.; Xu, Y. Dual-Emissive $CsPbBr_3$@Eu-BTC Composite for Self-Calibrating Temperature Sensing Application. *Cryst. Growth Des.* **2020**, *20*, 454–459. [CrossRef]
33. Zhou, X.; Li, S.; Zhou, X.; Chen, L.; Wu, J.; Yu, T.; Li, L.; Jiang, S.; Tang, X.; Xiang, G.; et al. Luminescent properties and ratiometric optical thermometry of Ln-BDC-F4 compounds. *Spectrochim. Acta A Mol. Biomol. Spectrosc.* **2020**, *224*, 117418. [CrossRef]
34. Zaręba, J.K.; Nyk, M.; Janczak, J.; Samoć, M. Three-Photon Absorption of Coordination Polymer Transforms UV-to-VIS Thermometry into NIR-to-VIS Thermometry. *ACS Appl. Mater. Interfaces* **2019**, *11*, 10435–10441. [CrossRef]
35. Chen, L.; Yu, H.; Li, Y.; Zhang, X.; Du, Y. Fabrication of a microporous Dy(III)-organic framework with polar channels for 5-Fu (fluorouracil) delivery and inhibiting human brain tumor cells. *Struct. Chem.* **2018**, *29*, 1885–1891. [CrossRef]
36. Zhang, Y.; Li, B.; Ma, H.; Zhang, L.; Jiang, H.; Song, H.; Zhang, L.; Luo, Y. A nanoscaled lanthanide metal–organic framework as a colorimetric fluorescence sensor for dipicolinic acid based on modulating energy transfer. *J. Mater. Chem. C* **2016**, *4*, 7294. [CrossRef]
37. Shu, Y.; Meng, Y.; Chen, M.-L.; Wang, J.-H. Isolation of hemoglobin with metal–organic frameworks $Y(BTC)(H_2O)_6$. *Chin. Chem. Lett.* **2015**, *26*, 1460–1464. [CrossRef]
38. Xiong, L.; Yu, L.; Li, S.; Feng, L.; Xiao, Y. Multifunctional lanthanide metal-organic framework based ratiometric fluorescence visual detection platform for alkaline phosphatase activity. *Microchim. Acta* **2021**, *188*, 236. [CrossRef]
39. Santos, P.F.; Ribeiro, J.; Luz, P.P. Physicochemical and electrochemical characterization of Ce/carbonaceous matrices-based composites. *Solid State Sci.* **2020**, *110*, 106479. [CrossRef]
40. Hu, M.; Shu, Y.; Kirillov, A.; Liu, W.; Yang, L.; Dou, W. Epoxy Functional Composites Based on Lanthanide Metal−Organic Frameworks for Luminescent Polymer Materials. *ACS Appl. Mater. Interfaces* **2021**, *13*, 7625–7634. [CrossRef]
41. Tong, Y.J.; Yu, L.D.; Zheng, J.; Liu, G.; Ye, Y.; Huang, S.; Chen, G.; Yang, H.; Wen, C.; Wei, S.; et al. Graphene Oxide-Supported Lanthanide Metal−Organic Frameworks with Boosted Stabilities and Detection Sensitivities. *Anal. Chem.* **2020**, *92*, 15550–15557. [CrossRef] [PubMed]
42. Moscoso, F.G.; Almeida, J.; Sousaraei, A.; Lopes-Costa, T.; Silva, A.M.G.; Cabanillas-Gonzalez, J.; Cunha-Silva, L.; Pedrosa, J.M. Luminescent MOF crystals embedded in PMMA/PDMS transparent films as effective NO_2 gas sensors. *Mol. Syst. Des. Eng.* **2020**, *5*, 1048–1056. [CrossRef]
43. Brunckova, H.; Mudra, E.; Rocha, L.; Nassar, E.; Nascimento, W.; Kolev, H.; Kovalcikova, A.; Molcanova, Z.; Podobova, M.; Medvecky, L. Preparation and characterization of isostructural lanthanide Eu/Gd/Tb metal-organic framework thin films for luminescent applications. *Appl. Surf. Sci.* **2021**, *542*, 148731. [CrossRef]
44. Gao, W.; Huang, H.; Zhou, A.-M.; Wei, H.; Liu, J.-P.; Zhang, X.-M. Three 3D LnIII-MOFs based on a nitro-functionalized biphenyltricarboxylate ligand: Syntheses, structures, and magnetic properties. *CrystEngComm* **2020**, *22*, 267–274. [CrossRef]
45. Dong, B.X.; Gua, X.J.; Xu, Q. Solvent effect on the construction of two microporous yttrium–organic frameworks with high thermostability viain situ ligand hydrolysis. *Dalton Trans.* **2010**, *39*, 5683–5687. [CrossRef]
46. Luo, J.; Xu, H.; Liu, Y.; Zhao, Y.; Daemen, L.L.; Brown, C.; Timofeeva, T.V.; Ma, S.; Zhou, H.-C. Hydrogen adsorption in a highly stable porous rare-earth metal-organic framework: Sorption properties and neutron diffraction studies. *J. Am. Chem. Soc.* **2008**, *130*, 9626–9627. [CrossRef] [PubMed]
47. Medina-Velazquez, D.Y.; Alejandre-Zuniga, B.Y.; Loera-Serna, S.; Ortiz, E.M.; Morales-Ramirez, A.D.J.; Garfias-Garcia, E.; Garcia-Murillo, A.; Falcony, C. An alkaline one-pot reaction to synthesize luminescent Eu-BTC MOF nanorods, highly pure and water-insoluble, under room conditions. *J. Nanopart. Res.* **2016**, *18*, 352. [CrossRef]
48. Zheng, Y.; Liu, K.; Sun, X.; Guan, R.; Su, H.; You, H.; Qi, C. A series of nano/micro-sized metal–organic frameworks with tunable photoluminescence properties. *Cryst. Eng. Comm.* **2015**, *17*, 2321–2326. [CrossRef]
49. Lo Presti, F.; Borzì, A.; Pellegrino, A.L.; Rossi, P.; Paoli, P.; Malandrino, G. Morphology controlled synthesis of yttrium metal–organic frameworks with a tritopic ligand. *Results Chem.* **2022**, *4*, 100640. [CrossRef]

50. Nikseresht, A.; Ghasemi, S.; Parak, S. [$Cu_3(BTC)_2$]: A metal–organic framework as an environment-friendly and economicall catalyst for the synthesis of tacrine analogues by Friedländer reaction under conventional and ultrasound irradiation. *Polyhedro* **2018**, *151*, 112–117. [CrossRef]
51. Tehrani, A.A.; Safarifard, V.; Morsali, A.; Bruno, G.; Rudbari, H.A. Ultrasound-assisted synthesis of metal–organic framewor nanorods of Zn-HKUST-1 and their templating effects for facile fabrication of zinc oxide nanorods via solid-state transformatior *Inorg. Chem. Commun.* **2015**, *59*, 41–45. [CrossRef]
52. Khan, N.A.; Jhung, S.H. Synthesis of metal-organic frameworks (MOFs) with microwave or ultrasound: Rapid reaction, phase selectivity, and size reduction. *Coord. Chem. Rev.* **2015**, *285*, 11–23. [CrossRef]
53. Kumari, P.; Kareem, A.; Jhariat, P.; Senthilkumar, S.; Panda, T. Phase Purity Regulated by Mechano-Chemical Synthesis o Metal−Organic Frameworks for the Electrocatalytic Oxygen Evolution Reaction. *Inorg. Chem.* **2023**, *62*, 3457–3463. [CrossRef [PubMed]
54. Liu, J.; Li, X.; Han, Y.; Wu, J.; Zhang, X.; Wang, Z.; Xu, Y. Synergetic Effect of Tetraethylammonium Bromide Addition on the Morpholog Evolution and Enhanced Photoluminescence of Rare-Earth Metal–Organic Frameworks. *Inorg. Chem.* **2020**, *59*, 14318–14325. [CrossRef
55. Xia, F.; Kang, S.; Xue, F.; Bu, X. Simple reductive synthesis of a novel mixed-lanthanide metal–organic framework with excellen cycling ability as a binder-free supercapacitor electrode. *Mater. Lett.* **2021**, *282*, 128715. [CrossRef]
56. Liu, Y.; Liu, B.; Zhou, Q.; Zhang, T.; Wu, W. Morphology effect of metal-organic framework HKUST-1 as a catalyst on benzen oxidation. *Chem. Res. Chin. Univ.* **2017**, *33*, 971–978. [CrossRef]
57. Xiao, J.D.; Qiu, L.G.; Ke, F.; Yuan, Y.P.; Xu, G.S.; Wang, Y.M.; Jiang, X. Rapid synthesis of nanoscale terbium-based metal–organi frameworks by a combined ultrasound-vapour phase diffusion method for highly selective sensing of picric acid. *J. Mater. Chem A* **2013**, *1*, 8745–8752. [CrossRef]
58. Hu, S.M.; Niu, H.L.; Qiu, L.G.; Yuan, Y.P.; Jiang, X.; Xie, A.J.; Shen, Y.H.; Zhu, J.F. Facile synthesis of highly luminescent nanowire of a terbium-based metal–organic framework by an ultrasonic-assisted method and their application as a luminescent probe fo selective sensing of organoamines. *Inorg. Chem. Commun.* **2012**, *17*, 147–150. [CrossRef]
59. Khan, N.A.; Haque, M.M.; Jhung, S.H. Accelerated Syntheses of Porous Isostructural Lanthanide–Benzenetricarboxylate (Ln–BTC) under Ultrasound at Room Temperature. *Eur. J. Inorg. Chem.* **2010**, *2010*, 4975–4981. [CrossRef]
60. Abbasi, A.R.; Rizvandi, M.; Azadbakht, A.; Rostamnia, S. Controlled uptake and release of imatinib from ultrasound nanoparticle $Cu_3(BTC)_2$ metal–organic framework in comparison with bulk structure. *J. Colloid Interface Sci.* **2016**, *471*, 112–117. [CrossRef]
61. Xu, W.; Li, G.; Lia, W.; Zhang, H. Facile room temperature synthesis of metal–organic frameworks from newly synthesize copper/zinc hydroxide and their application in adsorptive desulfurization. *RSC Adv.* **2016**, *6*, 37530–37534. [CrossRef]
62. Ley, S.V.; Low, C.M.R. Ultrasound in synthesis. In *Reactivity and Structure Concepts in Organic Chemistry*; Hafner, K., Lehn, J.-M. Rees, C.W., von Rague Schleyer, P., Trost, B.M., Zahradnik, R., Eds.; Springer-Verlag: Berlin/Heidelberg, Germany, 1989; Volum 27, pp. 1–129.
63. Daiguebonne, C.; Guilloa, O.; Gérault, Y.; Lecerf, A.; Boubekeur, K. Synthesis and crystal structure of two new rare earth trimesat complexes: $ErTMA(H_2O)_5{\cdot}3.5H_2O$ and $YTMA(H_2O)_6$. *Inorg. Chim. Acta* **1999**, *284*, 139–145. [CrossRef]
64. Binnemans, K. Interpretation of europium(III) spectra. *Coord. Chem. Rev.* **2015**, *295*, 1–45. [CrossRef]
65. Turro, N.J. *Modern Molecular Photochemistry*; University Science Books: Sausalito, CA, USA, 1991.
66. Liu, L.; Ding, R.; Mao, Y.; Sun, B. Theoretical investigations on the nitro-explosive sensing process of a MOF sensor: Roles o hydrogen bond and π-π stacking. *Chem. Phys. Lett.* **2022**, *793*, 139393. [CrossRef]
67. Sun, B.; Tao, T.; Liu, L.; Ding, R.; Mao, Y. Electron Transfer Facilitated by $\pi-\pi$ Stacking during the Nitrobenzene Recognitior Process of an MOF Sensor. *J. Phys. Chem. C* **2021**, *125*, 12433–12440. [CrossRef]
68. Du, Y.; Yang, H.; Liu, R.; Shao, C.; Yang, L. A multi-responsive chemosensor for highly sensitive and selective detection of Fe^{3+} Cu^{2+}, $Cr_2O_7^{2-}$ and nitrobenzene based on a luminescent lanthanide metal–organic framework. *Dalton Trans.* **2020**, *49*, 13003–13016 [CrossRef]
69. Chen, L.; Cheng, Z.; Peng, X.; Qiu, G.; Wang, L. Eu-Doped MOF-based high-efficiency fluorescent sensor for detecting 2,4 dinitrophenol and 2,4,6-trinitrophenol simultaneously. *Anal. Methods* **2022**, *14*, 44–51. [CrossRef]
70. Jornet-Mollá, V.; Dreessen, C.; Romero, F.M. Robust Lanthanoid Picolinate-Based Coordination Polymers for Luminescence and Sensing Applications. *Inorg. Chem.* **2021**, *60*, 10572–10584. [CrossRef]
71. Cheng, X.; Hu, J.; Li, J.; Zhang, M. Tunable emission and selective luminescence sensing for nitro- pollutants and metal ions based on bifunctional lanthanide metal-organic frameworks. *J. Lumin.* **2020**, *221*, 117100. [CrossRef]
72. Yang, X.; Liang, Y.; Shi, J.; Wang, Y.; Feng, W. Dual-lanthanide hollow luminescence metal-organic frameworks and their sensing performance for trace water and nitrobenzene. *Opt. Mater.* **2023**, *135*, 113343. [CrossRef]

molecules

Article

Anchoring of Polymer Loops on Enzyme-Immobilized Mesoporous ZIF-8 Enhances the Recognition Selectivity of Angiotensin-Converting Enzyme Inhibitory Peptides

Zefen Wang [1,2], Qian Zhou [1], Siyuan Liu [3], Dankui Liao [1,*], Pengru Liu [3,4] and Xiongdiao Lan [3,4,*]

1 School of Chemistry and Chemical Engineering, Guangxi University, Nanning 530004, China
2 Institute of Biological Manufacturing Technology Co., Ltd., Guangxi Institute of Industrial Technology, Nanning 530002, China
3 Guangxi Key Laboratory for Polysaccharide Materials and Modifications, Guangxi Minzu University, Nanning 530006, China
4 Key Laboratory of New Technology for Chemical and Biological Transformation Process of Guangxi Higher Education Institutes, Guangxi Minzu University, Nanning 530006, China
* Correspondence: liaodankuigx@163.com (D.L.); lanxiongdiao@163.com (X.L.)

Abstract: Immobilized angiotensin-converting enzyme (ACE) is a promising material for the rapid screening of antihypertensive drugs, but the nonspecific adsorption is a serious problem in separation processes involving complex biological products. In this study, triblock copolymers with dopamine (DA) block as anchors and PEG block as the main body (DA-PEGx-DA) were attached to an immobilized ACE (ACE@mZIF-8/PDA, AmZP) surface via the "grafting to" strategy which endowed them with anti-nonspecific adsorption. The influence of DA-PEGx-DA chain length on nonspecific adsorption was confirmed. The excellent specificity and reusability of the obtained ACE@mZIF-8/PDA/DA-PEG_{5000}-DA ($AmZPP_{5000}$) was validated by screening two known ACE inhibitory peptides Val-Pro-Pro (VPP, competitive inhibitory peptides of ACE) and Gly-Met-Lys-Cys-Ala-Phe (GF-6, noncompetitive inhibitory peptides of ACE) from a mixture containing active and inactive compounds. These results demonstrate that anchored polymer loops are effective for high-recognition selectivity and $AmZPP_{5000}$ is a promising compound for the efficient separation of ACE inhibitors in biological samples.

Keywords: polymer loops; angiotensin converting enzyme; immobilization of enzyme; mesoporous ZIF-8; recognition selectivity

1. Introduction

Recognized as a potential target for hypertension treatment, the angiotensin-converting enzyme (ACE) is the key enzyme for blood pressure regulation and is mainly localized on the membranes of lungs, heart, kidney, and intestinal cells [1–3]. The functional balance of ACE is crucial to blood pressure regulation. High ACE activity promotes angiotensin II production and bradykinin hydrolysis, leading to vasoconstriction and increased blood pressure [4,5]. Several chemical synthetic ACE inhibitors, such as captopril, enalaprilat, temocapril and lisinopril, have been used in hypertension treatment, but most of them produce side effects [6–8]. Thus, novel ACE inhibitors for developing anti-hypertension drugs are needed.

In recent years, numerous naturally-derived ACE inhibitory peptides have been reported to possess a strong inhibitory effect on ACE and have no negative side effects [9–11]. However, the screening of ACE inhibitory peptides from complex mixtures of natural products through traditional separation technologies is tedious and time-consuming. Thus, rapid screening methods for discovering ACE inhibitory peptides for novel hypertension therapeutics are in high demand.

Immobilized enzymes have been developed as affinity separation media for selectively enriching bioactive compounds from complex mixtures [12–15]. In recent decades, a number of materials, including magnetic beads [16–18], microspheres [19–21], capillary electrophoresis [22], and MOFs [23], have been used to immobilize ACE on surfaces and screen ACE inhibitory peptides from protein hydrolysate. However, these immobilized enzymes have a drawback: the natural conformation of an enzyme cannot be maintained during the separation process [24–26]. Recent studies on enzyme immobilization have demonstrated that enzymes can be encapsulated in situ in MOFs and retain the stability of their conformations [27–29]. However, MOFs-encapsulated enzymes have strong nonspecific adsorption for MOFs and thus show false selectivity [30,31]. Thus, anti-nonspecific protein adsorption is one of the main challenges in designing novel affinity separation media.

A common and promising method for solving this problem is surface modification with inert hydrophilic polymers [32–34]. Polymer constituents and morphology are important in this method. Polyethylene glycol (PEG) is one of the popular and most suitable polymer materials surface modification and considered the "gold standard" antifouling and biocompatible polymer by the FDA [35–37]. The highly-hydrated PEG chains on the surfaces of solids can effectively reduce nonspecific protein absorption. Most polymer chains are tethered through one end group to a surface. Loop-type polymers can be obtained by anchoring the chain ends of linear polymers to surfaces. Compared with linear polymers, loop-type polymers can better resist nonspecific protein adsorption [38–40]. In addition, introducing dopamine (DA) to the ends of polymers facilitates polymer surface grafting due to universal attachment and the ease of use of catechol groups [41].

In this work, ACE was encapsulated in situ in mesoporous ZIF-8 for the preparation of AmZP. A novel immobilized ACE AmZPP$_x$ (x = molecular weight of PEG) was obtained by anchoring an ABA triblock copolymer DA-PEG$_x$-DA with block A as a catechol anchor attracted to surfaces and block B in which PEG is a functional block on AmZP surfaces. The developed materials were systematically characterized. Two main inhibitory types of ACE inhibitory peptides are known: competitive and noncompetitive inhibition. The casein-derived peptide Val-Pro-Pro (VPP) is one of the ACE inhibitory peptides with competitive inhibition [42,43] and has been added to milk as a food additive. In our previous research, Gly-Met-Lys-Cys-Ala-Phe (GF-6) was isolated from *Saurida elongata* and acted as a noncompetitive inhibitor to ACE [44,45]. Thus, VPP and GF-6 were applied to investigate the adsorption performance, selectivity, separation ability, and reusability of AmZPP$_x$. This work is the first to apply loop-polymer-modified immobilized enzymes to bioactive compounds screening.

2. Results and Discussion

2.1. Synthesis of AmZPP$_x$

Highly selective materials for ACE inhibitory peptide screening were prepared using a polymer DA-assisted surface-modified ACE–MOF composite. The synthesis process is shown in Figure 1. First, ACE was encapsulated in situ in mZIF-8, approximately 90% of the enzyme was incorporated and the loading of protein was 0.11 mg/mg composites. Second, ACE@mZIF-8 was added to a DA solution water to form a PDA coating layer. This special structure of the AmZP afforded the materials low enzyme leakage [46]. Then, AmZP was added to a polymer solution for surface modification and reduction of nonspecific adsorption.

In this work, Bovine Serum Albumin (BSA) was used as a nonspecific protein model because it can easily attack many materials through nonspecific interactions [47]. Figure 2a shows the effect of the length of the PEG midblock on anti-nonspecific adsorption. The adsorption capacity of AmZPPx on BSA was significantly lower than of AmZP because of the repulsive force caused by the steric hindrance of loop-polymer-coated surfaces [40]. In addition, the BSA adsorption decreased with the increasing molecular weight of PEG because of the weak antiprotein effect when the polymer chain length was short. However, when the molecular weight of the PEG midblock increased to 5000, the effect on the adsorption of BSA was quite small. This result was consistent with that of a previous report [41]. Therefore, $AmZPP_{5000}$ was employed in subsequent studies. The effects of the dosages of DA-PEG_{5000}-DA and modification time on anti-nonspecific adsorption were surveyed (Figure 2b,c). BSA adsorption capacity decreased with increasing DA-PEG_{5000}-DA concentration and modification time possibly because of the increase in DA-PEG_{5000}-DA loading, and the strength of the steric hindrance effect increased. However, when the DA-PEG_{5000}-DA loading reached a certain value, anti-nonspecific performance did not change. Thus, the optimal synthesis conditions of $AmZPP_x$ were as follows: molecular weight of PEG midblock, 5000; DA-PEG_{5000}-DA concentration, 0.2 mM; and contact time, 30 min.

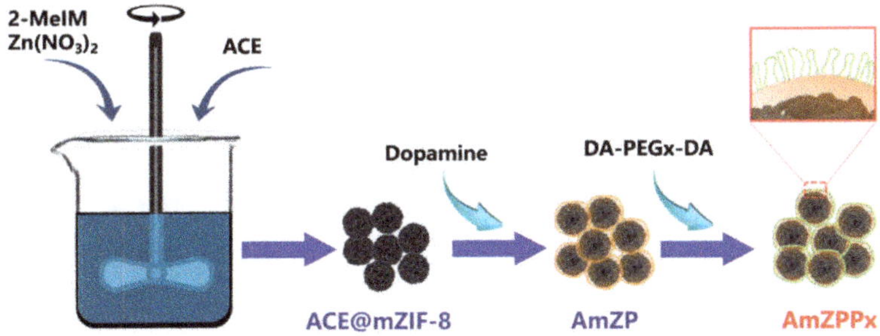

Figure 1. The synthesis procedure of $AmZPP_x$.

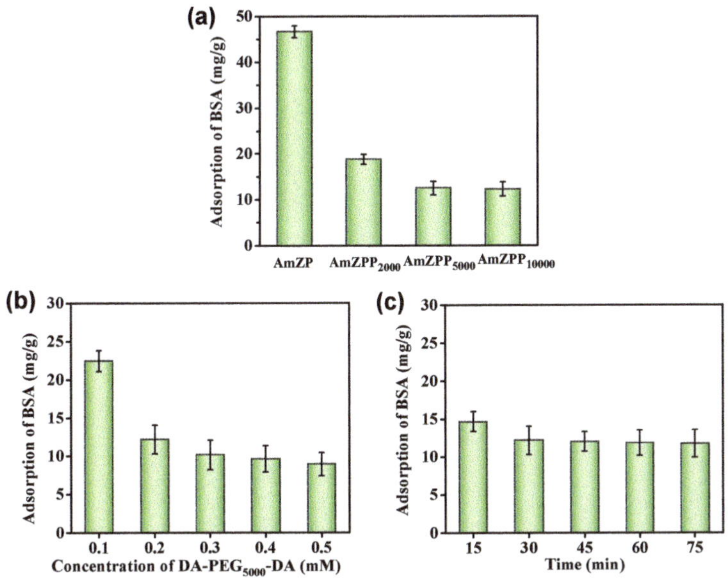

Figure 2. The BSA adsorption performance of $AmZPP_x$ with the different polymer length (**a**), concentration of DA-PEG_{5000}-DA (**b**) and contact time (**c**).

2.2. Characterization

The morphology of the synthesized materials was characterized by transmission electron microscopy (TEM) and scanning electron microscopy (SEM). As shown in Figure 3, AmZP and $AmZPP_{5000}$ had rough surfaces with diameters ranging from 100 nm to 200 nm, indicating that the polymer modification process did not significantly change the particle size. The TEM images indicated a distinct core–shell structure. Energy-dispersive spectroscopy (EDS) characterization of AmZP and $AmZPP_{5000}$ revealed the presence of ZIF-8

(Zn and N elements) and protein (S element), as shown in Figure 3e,f. The proportion of C element in AmZPP$_{5000}$ increased relative to that in AmZP, whereas the proportion of other elements decreased, illustrating successful modification by DA-PEG$_{5000}$-DA.

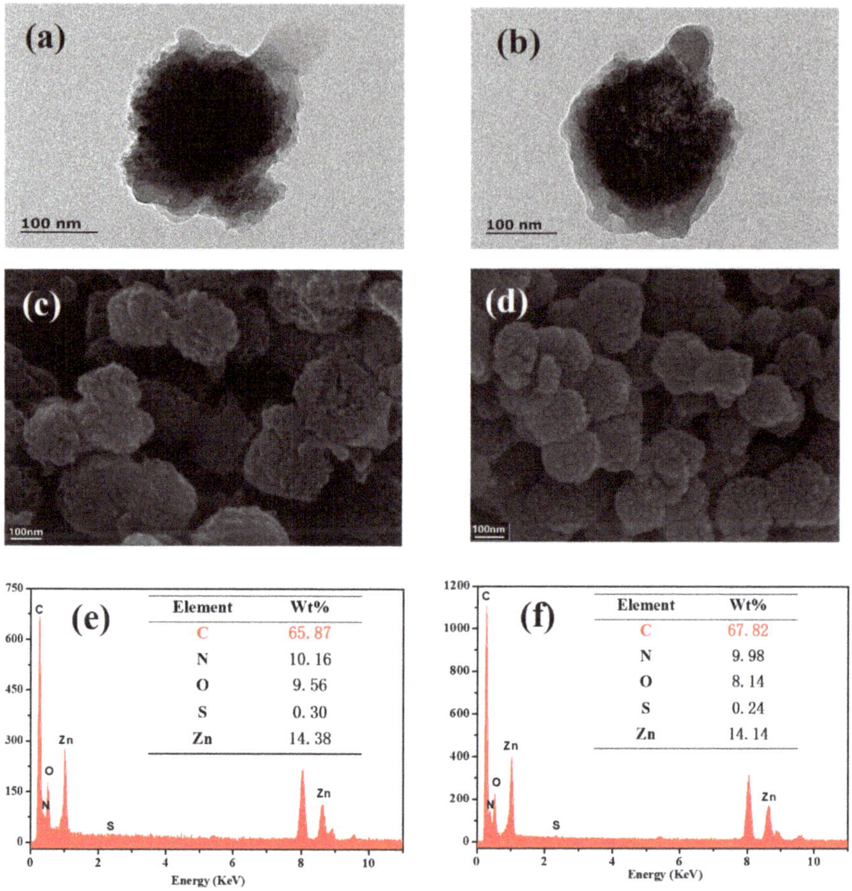

Figure 3. TEM, SEM and EDS images of AmZP (**a**, **c** and **e**, respectively) and AmZPP$_{5000}$ (**b**, **d** and **f**, respectively).

The Fourier-transform infrared (FT-IR) spectra of AmZP and AmZPP$_{5000}$ are shown in Figure 3. The wave numbers between 1600–1700 cm^{-1} and 1500–1580 cm^{-1} were attributed to amide I and amid II bands, respectively, and indicated the existence of enzymes [48]. The peak at ~1200 cm^{-1} represented the –OH vibration of phenol compounds, and indicated the presentation of PDA [46]. Moreover, a new absorption peak was detected at 841 cm^{-1} due to the bending vibration of –C–O–C–. This result indicated the successful combination of DA-PEG$_{5000}$-PEG.

The X-ray diffraction (XRD) patterns for AmZP and AmZPP$_{5000}$ are shown in Figure 4. Both materials presented the same characteristic peaks as ZIF-8 [48], but the peaks were diffused, indicating that the amorphous structures in the materials with short-range order and long-range disorder stacking. Mesopores in amorphous MOFs can make encapsulated enzymes have a higher activity than crystalline MOFs [49].

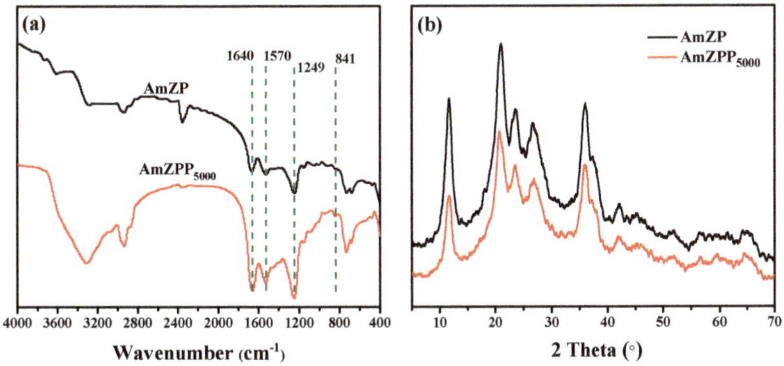

Figure 4. FT-IR (**a**) and XRD (**b**) spectra of AmZP and $AmZPP_{5000}$.

The specific surface areas and pore structures of AmZP and $AmZPP_{5000}$ were characterized by N_2 adsorption–desorption experiments. The Brunauer–Emmett–Teller (BET) curves displayed in Figure 5a exhibited a type IV adsorption isotherm with an H3 loop and indicated that both materials were mesoporous [50]. In addition, AmZP and $AmZPP_{5000}$ did not reach adsorption equilibrium when the relative pressure was close to the saturation pressure, indicating that the measured materials consisted of mesoporous slits. Based on the Barrett–Joyner–Halenda model (Figure 5b), the distribution diagram pore sizes of the two samples revealed an increase in 4–8 nm pores after DA-PEG_{5000}–PEG modification on AmZP, whereas the pore sizes were still mostly distributed in the 2–4 nm range. The specific surface area of $AmZPP_{5000}$ increased slightly from 51.16 m^2/g to 60.37 m^2/g compared with that of AmZP. These results suggested that DA-PEG_{5000}–PEG attached to AmZP surface via the "grafting to" strategy had no effect on the main structure of the material.

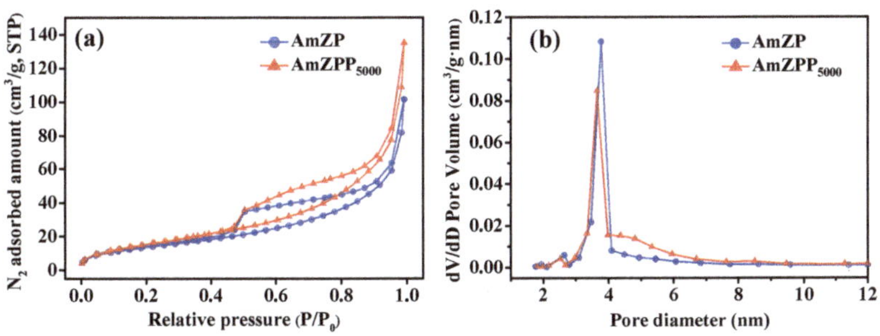

Figure 5. BET curves (**a**) and diagram pore sizes distribution (**b**) of AmZP and $AmZPP_{5000}$.

2.3. Specific Adsorption of $AmZPP_{5000}$

The specific adsorption analysis (Figure 6a) revealed that the adsorption capabilities of VPP and GF-6 first increased with peptide concentration and then remained unchanged after the maximum adsorption capability was reached. The mass transfer rate can be improved by increasing the concentration. Adsorption capacity is affected by the amount of encapsulated enzyme, cannot be increased indefinitely, and will not change after adsorption saturation is reached. Moreover, the adsorption capacity of GF-6 is higher than that of VPP because GF-6 is a noncompetitive inhibitor with multiple binding sites on ACE [45] and VPP is a competitive inhibitor that mainly binds to active sites of enzymes [43].

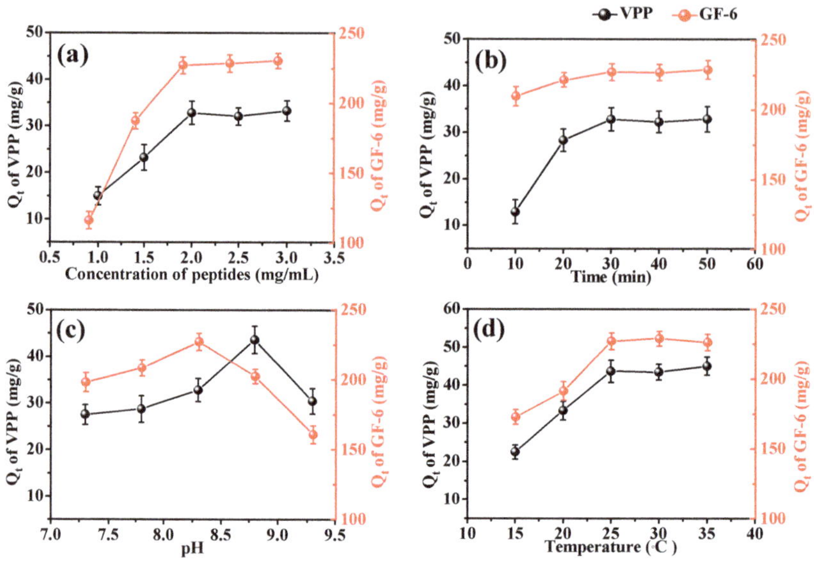

Figure 6. Effects of peptides concentration (**a**), time (**b**), pH (**c**) and temperature (**d**) on the adsorption of $AmZPP_{5000}$ on VPP and GF-6.

At initial VPP and GF-6 concentrations of 2 mg/mL, the relationship between the adsorption capacity of $AmZPP_{5000}$ and time was measured (Figure 6b). At 0–30 min, the adsorption capacity gradually increased with time because the peptides did not immediately reach the adsorption sites. At 30–60 min, the adsorption capacity remained unchanged, indicating that the binding sites of $AmZPP_{5000}$ reached saturation.

Figure 6c,d show the effects of pH and temperature on the adsorption of $AmZPP_{5000}$ on VPP and GF-6. The adsorption amount varied with pH and temperature because the adsorption of the two peptides was based on the specific interaction with ACE. The mass transfer rate and structure of the peptides change under experimental pH and temperature conditions, and thus change the adsorption behavior.

2.4. Adsorption Kinetics of $AmZPP_{5000}$ for VPP and GF-6

The pseudo-first order (PFO) and pseudo-second order (PSO) and intraparticle diffusion (IPD) models [51,52] were employed to fit the adsorption kinetic data obtained under the following conditions: temperature, 25 °C; pH, 8.3 (for GF-6) or 8.8 (for VPP); $AmZPP_{5000}$, 10 mg; peptides concentration, 2 mg/mL.

The fitting equation of the PFO model is expressed as follows:

$$Q_t = Q_e\left(1 - e^{-k_1 t}\right) \tag{1}$$

The fitting equation of the PSO model is expressed as follows:

$$Q_t = \frac{k_2 {Q_e}^2 t}{1 + k_2 Q_e t} \tag{2}$$

The fitting equation of the IPD model is expressed as follows:

$$Q_t = k_{id} t^{\frac{1}{2}} + C \tag{3}$$

where Q_t (mg/g) and Q_e (mg/g) are the adsorption capacities of peptides at a given time (t) and at equilibrium, respectively; k_1 and k_2 are the rate constants of the PFO and PSO kinetic models for adsorption, respectively; k_{id} is the intraparticle diffusion constants; and C is the effect of external boundary diffusion layer on adsorption process.

The fitting results are shown in Figure 7, and the adsorption parameters are displayed in Table 1. As expected, adsorption increased with contact time, and VPP and GF-6 achieved adsorption equilibrium after 30 min. The $Q_{e,cal}$ value obtained from the PFO model was closer to the experimental $Q_{e,exp}$ than the $Q_{e,cal}$ from the PSO model, and the R^2 in PFO was better than that in PSO. The adsorption rate constant of $AmZPP_{5000}$ for VPP (k_1 = 0.0774) was lower than that for GF-6 (k_1 = 0.2319), indicating a slower rate by the former than the latter. Above results suggested that the PFO model is more suitable for describing the adsorption processes of VPP and GF-6 on $AmZPP_{5000}$, which is related to physical adsorption.

The adsorption-rate-determining steps were investigated using the IPD model (Figure 7c,d and Table 2). The IPD fitting results indicated that the data points had three linear parts, suggesting the adsorption of VPP or GF-6 for $AmZPP_{5000}$ was associated with three continuous processes. The first step is related to surface diffusion; the second step is the diffusion process within a particle; and the third step is the dynamic equilibrium process of adsorption and desorption. Moreover, the plotted lines did not penetrate the origin ($C \neq 0$) and k_{id} in the second step was lower than that in the first step, indicating intraparticle diffusion was the main process but not the only control process.

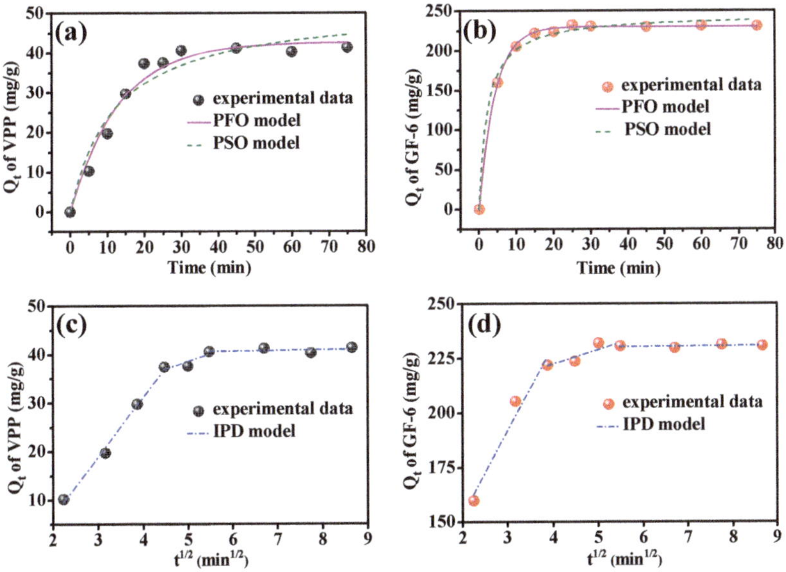

Figure 7. Adsorption kinetics of $AmZPP_{5000}$ for VPP (**a**,**c**) and GF-6 (**b**,**d**) fitting by PFO, PSO and IPD models.

Table 1. Kinetic parameters for VPP and GF-6 adsorption on $AmZPP_{5000}$.

		PFO Model			PSO Model		
	$Q_{e,exp}$ (mg/g)	$Q_{e,cal}$ (mg/g)	k_1	R^2	$Q_{e,cal}$ (mg/g)	k_2	R^2
VPP	40.88	42.68	0.0774	0.9757	51.61	0.00163	0.9472
GF-6	230.37	229.98	0.2319	0.9992	245.29	0.00182	0.9926

Table 2. IPD kinetic parameters for VPP and GF-6 adsorption on $AmZPP_{5000}$.

	First Step			Second Step			Third Step
	K_{id} (mg/(g·min$^{1/2}$)	C	R^2	K_{id} (mg/(g·min$^{1/2}$)	C	R^2	
VPP	12.26	−17.84	0.9924	3.134	22.91	0.8777	Balance stage
GF-6	38.49	76.70	0.9326	6.511	196.45	0.7069	

2.5. Adsorption Isotherm of $AmZPP_{5000}$ for VPP and GF-6

Experimental adsorption isotherm data were obtained under the following conditions: temperature, 25 °C; pH, 8.3 (for GF-6) or 8.8 (for VPP); AmZPP5000, 10 mg; contact time, 30 min. The Langmuir, Freundlich and D–R (Dubinin–Radushkevich) isotherm models [53] were used to describe the adsorption processes of VPP and GF-6.

The Langmuir model is represented by the following equation:

$$Q_e = \frac{Q_m k_L C_e}{1 + k_L C_e} \quad (4)$$

The Freundlich model is represented by the following equation:

$$Q_e = k_F C_e^{\frac{1}{n}} \quad (5)$$

The D–R adsorption model is represented by the following equation:

$$\ln Q_e = lnQ_m - B_{DR}\varepsilon_{DR}^2 \quad (6)$$

$$\varepsilon_{DR} = RTln\left(1 + \frac{1}{C_e}\right) \quad (7)$$

$$E = \frac{1}{\sqrt{2B_{DR}}} \quad (8)$$

where Q_e (mg/g) and Q_m (mg/g) are the equilibrium and theoretical maximum adsorption capacities of peptides, respectively; k_L represents the Langmuir constants; k_F and n represent the Freundlich constants; C_e is equilibrium peptide concentration; B_{DR} is the activity coefficient related to adsorption energy (mol^2/J^2); ε_{DR} is Polanyi potential energy; R is gas constant (8.314 J/mol/K) and T is absolute temperature (K); E is mean Gibbs free adsorption energy.

The results are shown in Figure 8 and Table 3. The adsorption of VPP and GF-6 on $AmZPP_{5000}$ was more consistent with the Langmuir model than with the Freundlich model, suggesting the adsorption was a monolayer process. Moreover, the maximum adsorption capacity Q_{max} of Langmuir fitting for VPP and GF-6 were 53.44 and 308.78 mg/g, respectively, indicating that $AmZPP_{5000}$ is a promising material for enrichment of ACE inhibitory peptides with excellent adsorption capacity.

The D–R model has no assumption of single-layer, homogeneous adsorption and could be used to distinguish physical adsorption from chemical adsorption. As seen from Figure 8c,d and Table 3, the D–R model fits the experiment data quite well. The E values of VPP and GF-6 were 0.552 and 0.561 kJ/mol, respectively. Usually, when $E < 8.0$ kJ/mol, physical adsorption can be considered as the main adsorption process, while chemical adsorption is the main adsorption process when $E > 8.0$ kJ/mol.

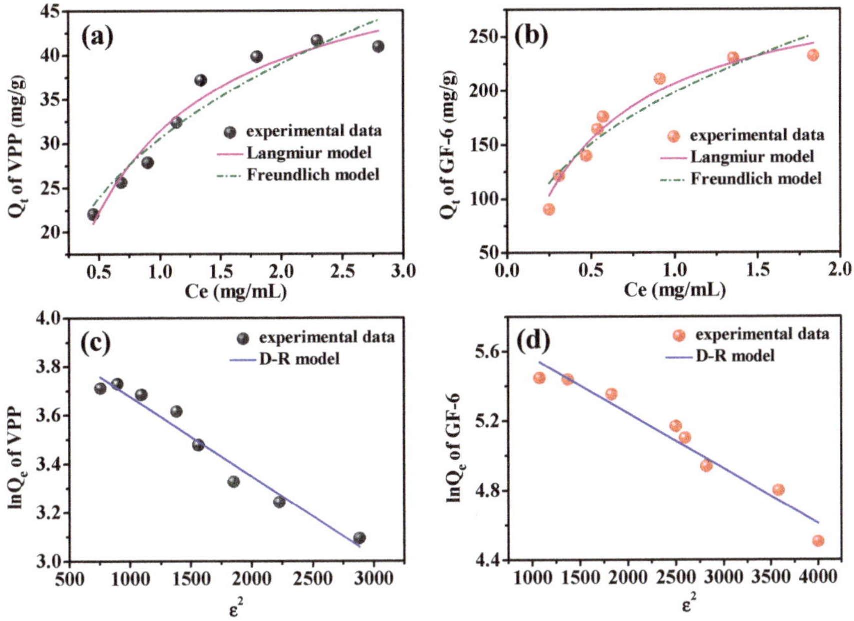

Figure 8. Adsorption isotherm of $AmZPP_{5000}$ for VPP (**a**,**c**) and GF-6 (**b**,**d**) fitting by Langmuir, Freundlich and D–R models.

Table 3. Adsorption isotherm parameters of $AmZPP_{5000}$ for VPP and GF-6.

	Langmuir Model			Freundlich Model			D–R Model		
	k_L (mL/mg)	Q_m (mg/g)	R^2	k_F	n	R^2	Q_m (mg/g)	E (KJ/mol)	R^2
VPP	1.4211	53.44	0.9501	30.53	2.816	0.9107	54.86	0.552	0.9556
GF-6	2.0131	308.78	0.9589	198.02	2.544	0.8860	357.70	0.561	0.9441

Above results indicated the adsorption process of VPP and GF-6 with $AmZPP_{5000}$ was physical monolayer adsorption with a limited number of isolated adsorption sites, which was consistent with previous reports that VPP and GF-6 bonded to ACE primarily via hydrogen bonding and hydrophobic interactions [43,45].

2.6. Desorption and Reusability of $AmZPP_{5000}$

Desorption and reusability are important indexes for materials' practical applications. The adsorption of ACE inhibitory peptides and ACE belongs to affinity adsorption, which usually requires a high concentration of NaCl for desorption [23]. As shown in Figure 9a,b, the desorption effect of 2 mol/L NaCl on VPP and GF-6 was significantly higher than 1.5, 1.0 and 0.5 mol/L and reached almost 80% at 1 h desorption time. This result suggested that 2 mol/L NaCl can be used as the desorption agent for the desorption of ACE inhibitory peptides from $AmZPP_{5000}$.

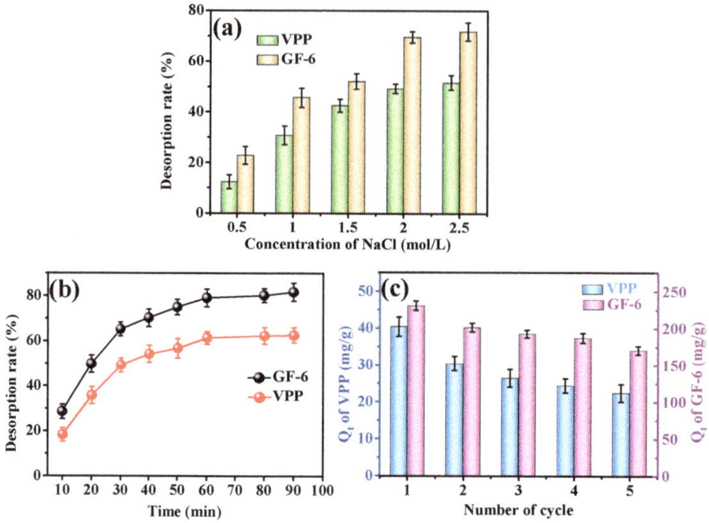

Figure 9. Desorption (**a**,**b**) and reusability (**c**) of $AmZPP_{5000}$.

It has been reported that amino acids and protein could destroy the structure of ZIF-8 because of the complexation of amino acids with zinc ions [54]. NaCl could also affect the structure of ZIF-8 [55]. As shown in Table 4, peptides NaCl and BSA could cause slight enzyme release from ACE@mZIF-8. The PDA layer could protect mZIF-8 against agents, making AmZP and AmZPP5000 more stable.

Table 4. Release of ACE from ACE@mZIF, AmZP and $AmZPP_{5000}$ after treatment with different agents for 1 h at 25 °C.

Agents	ACE@mZIF-8	Protein release (%) AmZP	$AmZPP_{5000}$
2 mg/mL VPP	7.32	1.37	1.24
2 mg/mL GF-6	6.21	1.08	1.13
		Enzyme activity release (%)	
2 mol/L NaCl	3.84	0.21	0.20
0.5 mg/mL BSA	1.52	0.00	0.00

The reusability of $AmZPP_{5000}$ was determined by repeating adsorption–desorption process. As shown in Figure 9c, $AmZPP_{5000}$ had excellent recycling performance after the fifth cycle, where the adsorption capacities of VPP and GF-6 were 22.39 mg/g and 170.58 mg/g, respectively. This decrease might have been caused by the incomplete desorption and the mass loss of the $AmZPP_{5000}$ after washing and during the repeated adsorption–desorption processes.

2.7. Recognition Selectivity of $AmZPP_{5000}$

A mixture containing VPP, GF-6, and BSA as the adsorption sample was used in investigating the recognition selectivity of AmZP, $AmZPP_{5000}$ and $mZPP_{5000}$ (without enzyme). The results are illustrated in Figure 10. Obviously, $AmZPP_{5000}$ and AmZP exhibited a higher adsorption capacity for VPP and GF-6 than $mZPP_{5000}$ for the specific interaction between peptides and ACE. All the materials displayed higher adsorption capacities for peptides than BSA. The reason was that BSA mainly bound to the materials' surfaces because of its diameter (~7 nm) [56], which was larger than the pore size of the materials (2–4 nm). BSA was unable to access the pores. By contrast, the small molecular peptides were transported to the mesopores and thus showed higher adsorption capacities. In the absence of anti-protein adsorption chain segments, the BSA adsorption capacities of $AmZPP_{5000}$ and $mZPP_{5000}$ was lower than that of AmZP. The results showed that the introduction of DA-PEG_{5000}-DA polymer loops can indeed improve the recognition selectivity of immobilized ACE.

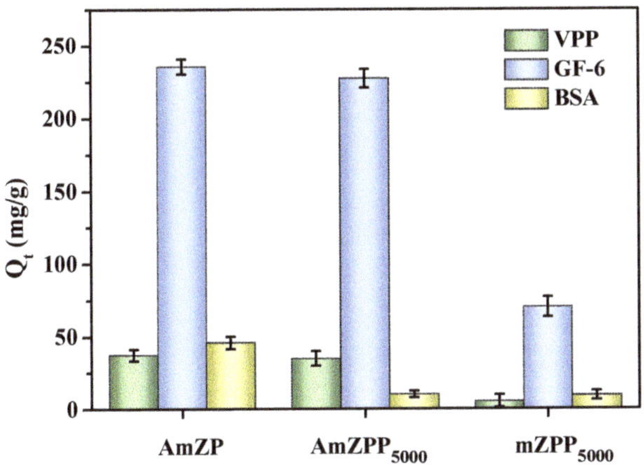

Figure 10. Adsorption data of $AmZPP_{5000}$ in mixed system.

2.8. Application of $AmZPP_{5000}$ in Hydrolysate

Based on the above results, $AmZPP_{5000}$ displayed excellent adsorption performance and reusability and provided a possibility for further application. The recognition selectivity of AmZP was compared with that of $AmZPP_{5000}$ by using a hydrolysate (obtained from *Leiognathus brevirostris*) containing ACE inhibitory peptides. The concentrations of proteins in the desorption solutions were determined using UV spectrophotometry. As shown in Figure 11 and Table 5, both materials had screening functions, and the desorbate of $AmZPP_{5000}$ displayed simpler components than that of AmZP and achieved lower IC_{50} value, which indicated a higher ACE inhibitory activity. These results demonstrated that the recognition selectivity of AmZP was successfully enhanced by anchoring polymer loops on the surface, and the $AmZPP_{5000}$ could be used to enriched ACE inhibitory peptides.

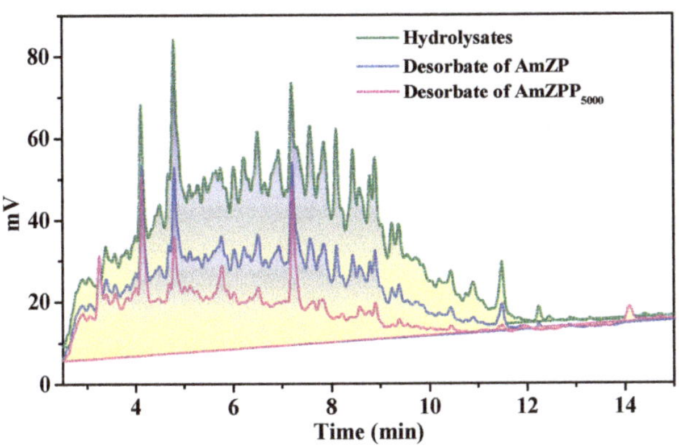

Figure 11. HPLC chromatogram on a Zorbax SB C18 column of samples. Experiments were carried out with a linear gradient of 5–50% acetonitrile in water (containing 0.1% TFA) over 15 min at a flow rate of 1 mL/min.

Table 5. Selective screening of ACE inhibitory peptides from hydrolysates by AmZP and $AmZPP_{5000}$.

Sample	IC_{50} (mg/mL)
Hydrolysates	0.623
Desorbate of AmZP	0.095
Desorbate of $AmZPP_{5000}$	0.034

3. Materials and Methods

3.1. Materials

Crude ACE was isolated according to the reported literature [44]. Hippuryl-histidyl-leucine (HHL; ≥99%) and ACE (from rabbit lung) were purchased from Sigma Chemical Co. (St. Louis, MO, USA). 2-MeIM (≥99%), DA (≥99%) and HPLC-grade trifluoroacetic acid were purchased from Shanghai Aladdin Biochemical Technology Co., Ltd. (Shanghai, China). DA-PEGx-DA (x = 2000, 5000 and 10,000) were purchased from Guangzhou Tanshtech Co., Ltd. (Guangzhou, China). VPP and GF-6 were obtained from GL Biochem (Shanghai) Ltd. (Shanghai, China). HPLC-grade methanol and acetonitrile were purchased from Shanghai Macklin Biochemical Co., Ltd. (Shanghai, China). All the other reagents were of analytical grade and purchased from XiLong Science (Guangdong, China).

3.2. Preparation of AmZP

For the synthesis of AmZP, a 100 mL beaker containing Zn $(NO3)_2 \cdot 6\,H_2O$ (0.0297 g), 2-MeIM (0.0657 g), 5 mL of crude ACE solution (protein concentration of 4 mg/mL and specific activity of 0.013 U/mg protein), and 30 mL of 0.1 M borate buffer solution (BBS, pH 7.3) was incubated at 30 °C for 30 min. The prepared ACE@mZIF-8 was collected by centrifugation at 8000 rpm for 10 min and washed with distilled water three times. Then, ACE@mZIF-8 was dispersed in 10 mL of 50 mM Tris-HCl (pH = 8.5), and 10 mg of DA was added. The mixture was gently agitated for 4 h at room temperature. The obtained AmZP was washed several times with deionized water and then vacuum freeze-dried.

3.3. Surface Modification with DA-PEG_x-DA

The prepared AmZP (100 mg) was dispersed in 20 mL of distilled water, and 20 mg of DA-PEG_x-DA (x = 2000, 5000, 10,000) was added. The mixture was mildly stirred for 1 h at room temperature. Then, the product $AmZPP_x$ was collected by centrifugation and washed with distilled water three times.

3.4. Characterization

SEM images were recorded using TESCAN MIRA LMS (Brno, Czech Republic) at a voltage of 3 kV. TEM images and EDS analysis results were obtained using JEOL JEM-F200 (Tokyo, Japan) at an operating voltage of 200 kV. FT-IR spectra were tested at a wavenumber range of 400–4000 cm^{-1} with a Nicolet IS10 (Waltham, MA, USA). XRD analysis was carried out using Mini Flex600 (Cu-Kα radiation), and the diffraction patterns were obtained from 5° to 70° with a scanning rate of 10°/min at 45 kV and 15 mA. Specific surface areas were measured at 77.35 K with the BET method in a relative pressure range (P/P_0) of 0.05–1. The amount of protein encapsulated in the material was determined by the Bradford method.

3.5. ACE Activity and Inhibitory Activity Assay

The digestion of HHL into hippuric acid (HA) and His-Leu (HL) using free or immobilized ACE was performed, and the activity of the enzyme was evaluated. Free ACE solution (protein concentration of 1 mg/mL) or immobilized ACE suspension (mass concentration of 5 mg/mL) were prepared in 0.1M BBS (containing 0.3 M NaCl, pH 8.3). Approximately 60 μL of the sample and 100 μL of 0.1M BBS (containing 0.3 M NaCl, pH 8.3) were incubated at 37 °C for 10 min. Then, 40 μL of 5 mM HHL (in BBS) was added and reacted for 15 min. The reaction was stopped by adding 100 μL of 1 M HCl. The activities of the free and immobilized ACE were measured by monitoring the released HA with the HPLC method.

ACE inhibitory activity was evaluated according to the above description with some modification. ACE solution (0.001 U/mL) was prepared using commercial ACE and the sample was used instead of the BBS buffer. The ACE inhibitory activity was calculated by the following equation [23]:

$$ACE\ inhibition\ rate = \frac{A_0 - A_1}{A_0} \times 100\% \quad (9)$$

where A_0 and A_1 are the peak areas of HA in the control group (without inhibitory peptides) and the sample (containing inhibitory peptides), respectively. IC_{50} is defined as the concentration of an inhibitor when half of the enzyme activity is suppressed.

3.6. Concentration of VPP and GF-6 Analysis

An RP-HPLC column (Zorbax SB-C18, 4.6 mm × 150 mm, Agilent, Palo Alto, CA, USA) was applied to measure the concentration of VPP and GF-6 with the solvent system consisting of 0.1% trifluoroacetic (in water, solvent A) and acetonitrile (containing 0.1% trifluoroacetic, solvent B). For VPP, a gradient of 5–15% solvent B over 15 min at a flow rate of 1 mL/min was used, while a gradient of 15–35% solvent B over 15 min at the same flow rate for GF-6. During the analysis process, the detector wavelength was set as 220 nm.

3.7. Specific/Nonspecific Adsorption Tests

BSA was used as a protein model for nonspecific adsorption. In this work, two known ACE inhibitory peptides, VPP (competitive inhibitory peptides of ACE) and GF-6 (noncompetitive inhibitory peptides of ACE), were selected for specific adsorption experiments.

The solutions of protein or peptides with different concentrations in 0.1 M BBS were prepared. For adsorption, the adsorbent and peptide solutions were mixed and shaken for a specified time. Then, the mixtures were centrifuged, and the concentrations of the peptides were detected via HPLC.

The adsorption amount of the materials was calculated as follows:

$$Q_t = \frac{(C_0 - C_t) \times V}{m} \quad (10)$$

$$Q_e = \frac{(C_0 - C_e) \times V}{m} \quad (11)$$

where Q_t and Q_e are the adsorption amounts of peptides (mg/g) at time t and at equilibrium, respectively; C_0, C, and *Ce* are the initial concentration, concentration at time t, and equilibrium concentration of a peptide solution (mg/mL), respectively; V is the volume of peptide solutions (mL); and m is the weight (g) of the immobilized enzyme.

3.8. Studies of Stability

The activity of ACE@mZIF-8, AmZP and $AmZPP_{5000}$ was monitored after incubation with 0.5 mg/mL BSA or 2 M NaCl at room temperature for 30 min. Because of the inhibitory effect of peptides on ACE, the stability of the material cannot be determined by measuring the activity after incubation with 2mg/mL VPP and GF-6. Therefore, the stability of the material was determined by measuring the protein concentration of the supernatant via the Bradford method.

3.9. Application of $AmZPP_{5000}$ in Hydrolysate Samples

The hydrolysates were prepared as follows: The fish meal of *Leiognathus brevirostris* (100 g) in 300 mL of 0.05 M PBS (pH 9.5) was heated in a boiling water bath for 10 min, and cooled to 45 °C. Alcalase was added in an enzyme/protein molar ratio of 4000 U/g for 3 h at 45 °C. Then, 1 mol/L NaOH was used to maintain the pH at 9.5 during hydrolysis. Enzymatic hydrolysis was terminated through inactivation in a boiling water bath for

10 min. The hydrolysate was collected by centrifugation (8000 rpm, 15 min) and then stored at −20 °C.

Approximately 10 mg of $AmZPP_{5000}$ was added to the hydrolysate with a shaking rate of 120 rpm for 30 min. After adsorption, the desorption experiment was performed in 2 M NaCl solution for 30 min at room temperature. The inhibitory activity of the supernatant was determined.

3.10. Statistical Analysis

All the experiments were carried out in triplicate, and the results are presented as average ± standard deviations. Statistically significant differences of the collected data were analyzed with one-way ANOVA. Statistical significance was established at $p < 0.05$.

4. Conclusions

In this work, the surface of immobilized ACE AmZP was successfully modified by ($DA\text{-}PEG_x\text{-}DA$) via a convenient "grafting to" strategy for the fabrication of a novel material ($AmZPP_x$) with high recognition selectivity for ACE inhibitory peptides. $AmZPP_{5000}$ with molecular weight of PEG midblock of 5000 showed the best anti-nonspecific adsorption performance. The specific adsorption capacities of $AmZPP_{5000}$ for VPP (competitive inhibitor for ACE) and GF-6 (noncompetitive inhibitor for ACE) were considerably higher than nonspecific adsorption capacity. Adsorption kinetic and adsorption isotherm studies showed that the adsorption of VPP and GF-6 belongs to physical monolayer adsorption and the diffusion was mainly controlled by intraparticle diffusion. Regarding the contribution of DA-PEGx-DA polymer loops, $AmZPP_{5000}$ presented excellent ability to specifically recognize ACE inhibitory peptides in actual hydrolysates.

Immobilized enzymes are becoming rapid and efficient platforms for screening bioactive components from complex products. We anticipate that capitalizing on the rational design of polymer loops modified on immobilized enzyme surfaces will benefit applications involving separation processes.

Author Contributions: Investigation, Data curation, Writing–original draft, Z.W.; Formal analysis and Validation, Q.Z. and S.L.; Funding acquisition, D.L., X.L. and P.L.; Writing–review and editing, Supervision and Project administration, D.L. and X.L. All authors have read and agreed to the published version of the manuscript.

Funding: This work was funded by the Natural Science Foundation of China (52162004), Guangxi Natural Science Foundation (2017GXNSFDA198052, 2021GXNSFAA075043 and 2022GXNSFBA035499), Nanning Excellent Youth Science and Technology Innovation and Entrepreneurship Talent Cultivation Project (RC20220106) and Guangxi Higher Education Institutes Young Teachers Basic Ability Improvement Project (2019KY0170).

Institutional Review Board Statement: Not applicable.

Informed Consent Statement: Not applicable.

Data Availability Statement: Not applicable.

Conflicts of Interest: The authors declare no conflict of interest.

Sample Availability: Samples of the compounds are available from the authors.

References

1. Xie, D.; Shen, Y.; Su, E.; Du, L.; Xie, J.; Wei, D. The effects of angiotensin I-converting enzyme inhibitory peptide vginyw and the hydrolysate of α-lactalbumin on blood pressure, oxidative stress and gut microbiota of spontaneously hypertensive rats. *Food Funct.* **2022**, *13*, 2743–2755. [CrossRef] [PubMed]
2. Bayannavar, P.K.; Kamble, R.R.; Joshi, S.D. Design and synthesis of angiotensin converting enzyme(ace)inhibitors: Analysis of the role of tetrazole ring appended to biphenyl moiety. *Chem. Sel.* **2022**, *7*, e202103336. [CrossRef]
3. Wu, N.; Zhao, Y.; Wang, Y.; Shuang, Q. Effects of ultra-high pressure treatment on angiotensin-converting enzyme (ace) inhibitory activity, antioxidant activity, and physicochemical properties of milk fermented with lactobacillus delbrueckii qs306. *J. Dairy Sci.* **2022**, *105*, 1837–1847. [CrossRef] [PubMed]

4. Mehmood, A.; Pan, F.; Ai, X.; Tang, X.; Cai, S.; Soliman, M.M.; Yanli, N. Novel angiotensin-converting enzyme (ace) inhibitory mechanism of peptides from macadamia integrifolia antimicrobial protein 2 (miamp2). *J. Food Biochem.* **2022**, *46*, e14168. [CrossRef]
5. Dong, J.; Wang, S.; Yin, X.; Fang, M.; Gong, Z.; Wu, Y. Angiotensin i converting enzyme(ace)inhibitory activity and antihypertensive effects of rice peptides. *Food Sci. Hum. Well.* **2022**, *11*, 1539–1543. [CrossRef]
6. Noguchi, Y.; Murayama, A.; Esaki, H.; Sugioka, M.; Koyama, A.; Tachi, T.; Teramachi, H. Angioedema caused by drugs that prevent the degradation of vasoactive peptides: A pharmacovigilance database study. *J. Clin. Med.* **2021**, *10*, 5507.
7. Kaya, A.; Tatlisu, M.A.; Kaya, T.K.; Yildirimturk, O.; Gungor, B.; Karatas, B. Sublingual vs. oral captopril in hypertensive crisis. *J. Emerg. Med.* **2016**, *50*, 108–115. [CrossRef] [PubMed]
8. Syed, S.B.; Reyes, J.V.M.; Baig, M. Lisinopril-induced Acute Necrotizing Pancreatitis. *Cureus* **2021**, *13*, e14642. [CrossRef] [PubMed]
9. Solanki, D.; Sakure, A.; Prakash, S. Characterization of angiotensin I-converting enzyme (ACE) inhibitory peptides produced in fermented camel milk (Indian breed) by Lactobacillus acidophilus NCDC-15. *J. Food Sci. Technol.* **2022**, *59*, 3567–3577.
10. Martinez-Sanchez, S.M.; Gabaldon-Hernandez, A.; Montoro-Garcia, S. Unravelling the molecular mechanisms associated with the role of food-derived bioactive peptides in promoting cardiovascular health. *J. Funct. Foods.* **2020**, *64*, 103645. [CrossRef]
11. Shanmugam, V.P.; Kapila, S.; Kemgang, T.S.; Reddi, S.; Kapila, R.; Muthukumar, S.; Rajesh, D. Isolation and Characterization of Angiotensin Converting Enzyme Inhibitory Peptide from Buffalo Casein. *Int. J. Pept. Res. Ther.* **2021**, *27*, 1481–1491. [CrossRef]
12. Li, X.; Su, Z.; Luo, Y.; Chen, X.; Luo, J.; Pinelo, M. Modelling of oligodextran production via an immobilized enzyme membrane reactor: Bioreaction-separation coupling mechanism. *Sep. Purif. Technol.* **2021**, *282*, 120024. [CrossRef]
13. Mujtaba, N.; Jahan, N.; Sultana, B.; Zia, M.A. Isolation and characterization of antihypertensive peptides from soy bean protein. *Braz. J. Pharm. Sci.* **2021**, *57*, e19061. [CrossRef]
14. Chuang, H.; Huang, C.; Hung, T.; Huang, L.; Lin, N. Development of biotinylated and magnetic bead-immobilized enzymes for efficient glyco-engineering and isolation of antibodies. *Bioorg. Chem.* **2021**, *112*, 104863. [CrossRef]
15. Wan, G.; Ma, X.; Jin, L.; Chen, J. α-glucosidase immobilization on magnetic core-shell metal-organic frameworks for inhibitor screening from traditional chinese medicines. *Colloid. Surf. B* **2021**, *205*, 111847. [CrossRef] [PubMed]
16. Tang, W.; Jia, B.; Zhou, J.; Liu, J.; Wang, J.; Ma, D.; Li, P.; Chen, J. A method using angiotensin converting enzyme immobilized on magnetic beads for inhibitor screening. *J. Pharmaceut. Biomed.* **2019**, *164*, 223–230. [CrossRef]
17. Wei, N.; Zhao, J.; Wu, G.; Cao, W.; Luo, P.; Zhang, Z.; Chen, G.; Wen, L. Rapid Screening and Identification of Antitumor Ingredients from the Mangrove Endophytic Fungus Using an Enzyme-Immobilized Magnetic Nanoparticulate System. *Molecules* **2021**, *26*, 2255. [CrossRef]
18. Wu, X.; Qiu, B.; Chen, Y.; Shi, Y.; Zhu, J.; Liu, X.; Zhao, D. Online coupling Fe3O4@ZIF-67@alpha-glucosidase biomicroreactor with high performance liquid chromatography for rapid screening of alpha-glucosidase inhibitors in tea and their inhibitory activity research. *J. Chromatogr. B.* **2020**, *1159*, 122398. [CrossRef]
19. Megías, C.; Pedroche, J.; Yust, M.; Alaiz, M.; Girón-Calle, J.; Millán, F.; Vioque, J. Immobilization of angiotensin-converting enzyme on glyoxyl-agarose. *J. Agric. Food Chem.* **2006**, *54*, 4641–4645. [CrossRef]
20. Hong, Y.; Li, T.; Chen, B.; Yao, S. Immobilization of angiotensin converting-enzyme on chitosan microspheres. *Chem. J. Chin. Univ.* **2009**, *30*, 328–331.
21. Hui, L.; Yin, Q.; Jin, Q.; Xinyu, L. A novel strategy for screening angiotensin-converting enzyme inhibitors from natural products based on enzyme-immobilized ligand fishing combined with active-site blocking and directional enrichment. *J. Chromatogr. B.* **2022**, *1195*, 123203.
22. Meizhong, T.; Jingwu, K. Enzyme inhibitor screening by capillary electrophoresis with an on-column immobilized enzyme microreactor created by an ionic binding technique. *Anal. Chem.* **2006**, *78*, 2514–2520.
23. Feng, X.; Liao, D.; Sun, L.; Wu, S.; Lan, P.; Wang, Z.; Li, C.; Zhou, Q.; Lu, Y.; Lan, X. Affinity Purification of Angiotensin Converting Enzyme Inhibitory Peptides from Wakame (Undaria Pinnatifida) Using Immobilized ACE on Magnetic Metal Organic Frameworks. *Mar. Drugs.* **2021**, *19*, 177. [CrossRef] [PubMed]
24. Altinkaynak, C.; Tavlasoglu, S.; Ÿzdemir, N.; Ocsoy, I. A new generation approach in enzyme immobilization: Organic-inorganic hybrid nanoflowers with enhanced catalytic activity and stability. *Enzyme Microb. Tech.* **2016**, *93–94*, 105–112. [CrossRef] [PubMed]
25. Lian, X.; Fang, Y.; Joseph, E.; Wang, Q.; Li, J.; Banerjee, S.; Wang, X.; Zhou, H. Enzyme-mof (metal-organic framework) composites. *Chem. Soc. Rev.* **2017**, *46*, 3386–3401. [CrossRef]
26. Fried, D.; Brieler, F.; Frba, M. Designing inorganic porous materials for enzyme adsorption and applications in biocatalysis. *ChemCatChem* **2013**, *5*, 862–884. [CrossRef]
27. Liang, W.; Xu, H.; Carraro, F.; Maddigan, N.K.; Li, Q.; Bell, S.G.; Huang, D.; Tarzia, A.; Solomon, M.B.; Amenitsch, H. Enhanced activity of enzymes encapsulated in hydrophilic metal-organic frameworks. *J. Am. Chem. Soc.* **2019**, *141*, 2348–2355. [CrossRef]
28. Li, Y.; Yuan, J.; Ren, H.; Ji, C.; Tao, Y.; Wu, Y.; Zhang, Y.; Cheng, L. Fine-tuning the micro-environment to optimize the catalytic activity of enzymes immobilized in multivariate metal-organic frameworks. *J. Am. Chem. Soc.* **2021**, *143*, 15378–15390. [CrossRef]
29. Chen, Y.; Jimenez-Angeles, F.; Qiao, B.; Krzyaniak, M.D.; Farha, O.K. Insights into the enhanced catalytic activity of cytochrome c when encapsulated in a metal-organic framework. *J. Am. Chem. Soc.* **2020**, *142*, 18579–18582. [CrossRef]
30. Peng, S.; Liu, J.; Qin, Y.; Wang, H.; Yu, X. Metal-organic framework encapsulating hemoglobin as a high-stable and long-circulating oxygen carriers to treat hemorrhagic shock. *ACS Appl. Mater. Interfaces* **2019**, *11*, 35604–35612. [CrossRef]

31. Qian, L.; Liu, W.; Liu, H.; Nica, V.; Zhang, S.; Zhou, Q.; Song, W.; Zhang, Q. Fabrication of raspberry-like cytochrome c surface imprinted nanoparticles based on mof composites for high-performance protein separation. *ACS Appl. Mater. Interfaces* **2021**, *1*
31010–31020. [CrossRef]
32. Yang, Z.; Zhang, X.; Xie, M.; Wu, H.; Yoshioka, T.; Saeki, D.; Matsuyama, H. Antifouling thin-film composite membranes with multi-defense properties by controllably constructing amphiphilic diblock copolymer brush layer. *J. Membr. Sci.* **2020**, *614*, 11851 [CrossRef]
33. Niu, J.; Wang, H.; Chen, J.; Chen, X.; Han, X.; Liu, H. Bio-inspired zwitterionic copolymers for antifouling surface and oil-wate separation. *Colloid. Surf. A* **2021**, *626*, 127016. [CrossRef]
34. He, Y.; Wan, X.; Lin, W.; Li, J.; Li, Z.; Luo, F.; Li, J.; Tan, H.; Fu, Q. The synergistic effect of hierarchical structure and alkyl chai length on the antifouling and bactericidal properties of cationic/zwitterionic block polymer brushes. *Biomater. Sci.* **2020**, *8* 6890–6902. [CrossRef]
35. Kanamaru, T.; Sakurai, K.; Envelope, S.F. Impact of polyethylene glycol (peg) conformations on the in vivo fate and drug releas behavior of pegylated core-cross-linked polymeric nanoparticles. *Biomacromolecules* **2022**, *23*, 3909–3918. [CrossRef] [PubMed]
36. Li, M.Y.; Jiang, S.; Simon, J.; Passlick, D.; Frey, M.L.; Wagner, M.; Mailander, V.; Crespy, D.; Landfester, K. Brush conformatio of polyethylene glycol determines the stealth effect of nanocarriers in the low protein adsorption regime. *Nano Lett.* **2021**, *21* 1591–1598. [CrossRef] [PubMed]
37. Yoo, J.; Kim, K.; Kim, S.; Park, H.; Shin, H.; Joo, J. Tailored polyethylene glycol grafting on porous nanoparticles for enhance targeting and intracellular sirna delivery. *Nanoscale* **2022**, *14*, 14482–14490. [CrossRef] [PubMed]
38. Kang, T.; Banquy, X.; Heo, J.H.; Lim, C.N.; Lynd, N.A.; Lundberg, P.; Oh, D.X.; Lee, H.K.; Hong, Y.K.; Hwang, D.S.; et a Mussel-inspired anchoring of polymer loops that provide superior surface lubrication and antifouling properties. *Acs Nano.* **2016** *10*, 930–937. [CrossRef]
39. Wang, H.; Wang, H.; Zhang, Z.; Zhang, Z.; Chen, J.; Chen, J. Conformation-dominated surface antifouling and aqueous lubricatio *Colloid. Surf. B* **2022**, *214*, 112452. [CrossRef]
40. Xiong, C.; Xiong, W.; Mu, Y.; Pei, D.; Wan, X. Mussel-inspired polymeric coatings with the antifouling efficacy controlled b topologies. *J. Mater. Chem. B* **2022**, *10*, 9295–92304. [CrossRef]
41. Shin, E.; Lim, C.; Kang, U.J.; Kim, M.; Park, J.; Kim, D.; Choi, W.; Hong, J.; Baig, C.; Lee, D.W. Mussel-inspired copolyether loo with superior antifouling behavior. *Macromolecules* **2020**, *53*, 3551–3562. [CrossRef]
42. Pina, A.S.; Roque, A.C.A. Studies on the molecular recognition between bioactive peptides and angiotensin-converting enzym *J. Mol. Recognit.* **2010**, *22*, 162–168. [CrossRef] [PubMed]
43. Sawada, Y.; Sakamoto, Y.; Toh, M.; Ohara, N.; Hatanaka, Y.; Naka, A.; Kishimoto, Y.; Kondo, K.; Iida, K. Milk-derived peptid val-pro-pro (VPP) inhibits obesity-induced adipose inflammation via an angiotensin-converting enzyme (ACE) dependen cascade. *Mol. Nutr. Food Res.* **2015**, *59*, 2502–2510. [CrossRef] [PubMed]
44. Lan, X.; Liao, D.; Wu, S.; Wang, F.; Sun, J.; Tong, Z. Rapid purification and characterization of angiotensin converting enzym inhibitory peptides from lizard fish protein hydrolysates with magnetic affinity separation. *Food Chem.* **2015**, *182*, 136–142 [CrossRef] [PubMed]
45. Lan, X.; Sun, L.; Muhammad, Y.; Wang, Z.; Liu, H.; Sun, J.; Zhou, L.; Feng, X.; Liao, D.; Wang, S. Studies on the interactio between angiotensin-converting enzyme (ace) and ace inhibitory peptide from saurida elongata. *J. Agric. Food Chem.* **2018**, *66* 13414–13422. [CrossRef] [PubMed]
46. Zhang, S.; Shi, J.; Zhang, Y.; Wu, Y.; Chen, Y.; Messersmith, P.B.; Jiang, Z. Polymer@mofs capsules prepared through controlled interfacial mineralization for switching on/off enzymatic reactions. *Appl. Mater. Today.* **2018**, *13*, 320–328. [CrossRef]
47. Rashid, S.; Ward-Bond, J.; Krupin, O.; Berini, P. Non-specific adsorption of protein to microfluidic materials. *Colloid. Surf. B* **2021** *208*, 112138. [CrossRef]
48. Hu, Y.L.; Dai, L.; Liu, D.; Du, W. Hydrophobic pore space constituted in macroporous ZIF-8 for lipase immobilization greatl improving lipase catalytic performance in biodiesel preparation. *Biotechnol. Biofuels.* **2020**, *13*, 86. [CrossRef]
49. Wu, X.; Yue, H.; Zhang, Y.; Gao, X.; Li, X.; Wang, L.; Cao, Y.; Hou, M.; An, H.; Zhang, L.; et al. Packaging and delivering enzyme by amorphous metal-organic frameworks. *Nat. Commun.* **2019**, *10*, 5165. [CrossRef]
50. Dou, D.; Wei, D.; Guan, X.; Liang, Z.; Lan, L.; Lan, X.; Liu, P.; Mo, H.; Lan, P. Adsorption of copper (II) and cadmium (II) ions b in situ doped nano-calcium carbonate high-intensity chitin hydrogels. *J. Hazard. Mater.* **2021**, *423*, 127137. [CrossRef]
51. Georgianos, P.; Pournara, A.D.; Andreou, E.K.; Armatas, G.S.; Manos, M.J. Composite Materials Based on a Zr^{4+} MOF and Aluminosilicates for the Simultaneous Removal of Cationic and Anionic Dyes from Aqueous Media. *Molecules* **2023**, *28*, 815 [CrossRef] [PubMed]
52. Simonin, J.P.; Bouté, J. Intraparticle diffusion-adsorption model to describe liquid/solid adsorption kinetics. *Rev. Mex. Ing. Quim* **2016**, *15*, 161–173.
53. Tourani, S.; Behvandi, A. Synthesis of MIL-101(Cr)/Sulfasalazine (Cr-TA@SSZ) hybrid and its use as a novel adsorbent fo adsorptive removal of organic pollutants from wastewaters. *J. Porous Mat.* **2022**, *29*, 1441–1462. [CrossRef]
54. Gao, Y.; Doherty, C.M.; Mulet, X. A Systematic Study of the Stability of Enzyme/Zeolitic Imidazolate FrameworkComposites in Various Biologically Relevant Solutions. *Chem. Sel.* **2020**, *5*, 13766–13774.

55. Shujin, P.; Xuan, Z.; Yang, C.; NAseer, S.; Zhang, X.; Ouyang, J.; Li, D. The effects of NaCl on enzyme encapsulation by zeolitic imidazolate frameworks-8. *Enzyme Microb. Tech.* **2019**, *122*, 1–6.
56. Wang, S.; Chen, K.; Li, L.; Guo, X. Binding between proteins and cationic spherical polyelectrolyte brushes: Effect of pH, ionic strength, and stoichiometry. *Biomacromolecules* **2013**, *14*, 818–827. [CrossRef]

molecules

MDPI

Article

Metal and Ligand Effect on the Structural Diversity of Divalent Coordination Polymers with Mixed Ligands: Evaluation for Photodegradation

Manivannan Govindaraj [1], Shih-Ying Zhong [1], Chia-Her Lin [2,*] and Jhy-Der Chen [1,*]

[1] Department of Chemistry, Chung-Yuan Christian University, Chung Li, Taoyuan City 320, Taiwan
[2] Department of Chemistry, National Taiwan Normal University, Taipei 106, Taiwan
* Correspondence: chiaher@ntnu.edu.tw (C.-H.L.); jdchen@cycu.edu.tw (J.-D.C.); Tel.: +886-3-265-3351 (J.-D. C.)

Abstract: Eight coordination polymers constructed from divalent metal salts, *N,N′*-bis(pyridin-3-ylmethyl)terephthalamide (**L**), and various dicarboxylic acids are reported, affording $[Co(\mathbf{L})(5\text{-ter-}IPA)(H_2O)_2]_n$ (5-tert-H_2IPA = 5-tert-butylisophthalic acid), **1**, $\{[Co(\mathbf{L})(5\text{-}NO_2\text{-}IPA)]\cdot 2H_2O\}_n$ (5-NO_2-H_2IPA = 5-nitroisophthalic acid), **2**, $\{[Co(\mathbf{L})_{0.5}(5\text{-}NH_2\text{-}IPA)]\cdot MeOH\}_n$ (5-NH_2-H_2IPA = 5-aminoisophthalic acid), **3**, $\{[Co(\mathbf{L})(MBA)]\cdot 2H_2O\}_n$ (H_2MBA = diphenylmethane-4,4′–dicarboxylic acid), **4**, $\{[Co(\mathbf{L})(SDA)]\cdot H_2O\}_n$ (H_2SDA = 4,4-sulfonyldibenzoic acid), **5**, $\{[Co_2(\mathbf{L})_2(1,4\text{-}NDC)_2(H_2O)_2]\cdot 5H_2O\}_n$ (1,4-H_2NDC = naphthalene-1,4-dicarboxylic acid), **6**, $\{[Cd(\mathbf{L})(1,4\text{-}NDC)(H_2O)]\cdot 2H_2O\}_n$, **7**, and $\{[Zn_2(\mathbf{L})_2(1,4\text{-}NDC)_2]\cdot 2H_2O\}_n$, **8**, which were structurally characterized by using single-crystal X-ray diffraction. The structural types of **1–8** are subject to the metal and ligand identities, showing a 2D layer with the **hcb**, a 3D framework with the **pcu**, a 2D layer with the **sql**, a polycatenation of 2-fold interpenetrated 2D layer with the **sql**, a 2-fold interpenetrated 2D layer with the 2,6L1, a 3D framework with the **cds**, a 2D layer with the 2,4L1, and a 2D layer with the $(10^2\cdot 12)(10)_2(4\cdot 10\cdot 12^4)(4)$ topologies, respectively. The investigation on the photodegradation of methylene blue (MB) by using complexes **1–3** reveals that the degradation efficiency may increase with increasing surface areas.

Keywords: coordination polymer; crystal structure analysis; dicarboxylate; coordination mode

1. Introduction

Coordination polymers (CPs) have been intensively investigated by scientists in recent years because of their intriguing architectures and prospective applications in magnetism, luminescence, catalysis, gas storage, and sensing [1–4]. The coordination of spacer ligands to metal ions during the self-assembly process may result in the production of infinite one-dimensional (1D), two-dimensional (2D), or three-dimensional (3D) CPs, which are subject to the careful selection of metal ion and spacer ligands with diverse functionalities and flexibility. Despite the fact that many fascinating CPs have been reported, control of the structural variety remains a difficulty in the field of crystal engineering, and the factors that influence the structural diversity are less well understood [5,6].

The mixed ligand assembly technique has been employed to develop novel CPs [7]. In this context, mixed-ligand techniques including two distinct types of ligands with unique functions, such as polycarboxylate paired with a bis-pyridyl-bis-amide (bpba)-based N donor ligand, have been introduced as an effective way for adjusting structural diversity in CPs. Bpba ligands are remarkable ligands that may be modified to yield intriguing CPs [8], due to the fact that the majority of bpba ligands are flexible; however, others are semi-rigid.

Herein, we adopted the semi-rigid *N,N′*-bis(pyridin-3-ylmethyl)terephthalamide (**L**), as shown in Figure 1, and differently substituted dicarboxylic acid as part of our ongoing research into understanding the relationship between the mixed-ligand system and the structural variety of the new CPs. Several CPs containing **L** and dicarboxylate ligands have been structurally characterized. Three CPs, namely $[Co(\mathbf{L})(HIPA)(H_2O)_2]\cdot H_2O$

(H_2HIPA = 5-Hydroxyisophthalic acid), [Ni(**L**)(HIPA)(H_2O)$_2$]·H_2O, and [Cu(**L**)(HIPA)]·4H_2O, show similar 2D cellular networks for the first two complexes, and the latter has a 4-connected 3D structure with a semicircle 1D channel [9]. On the other hand, [Zn(**L**)(1,4-BDC)]·H_2O (1,4-H_2BDC = terephthalic acid), [Zn(**L**)(1,3- BDC)]·H_2O (1,3-H_2BDC = isophthalic acid), [Zn(**L**)(1,2-BDC)] (1,2-H_2BDC = phthalate), and [Cd(**L**)$_{0.5}$(1,2-BDC)(H_2O)] display 2D structures with 6^3 and 4^4 topologies and a 4-connected 3D framework and a 1D structure, respectively [10], whereas [Cd_3(**L**)$_2$(1,4-bdc)$_3$]·4H_2O and [Cd(**L**)(1,4-bdc)]·2H_2O are (3,5)-connected nets with the $(3{\cdot}7^2)(3^2{\cdot}4{\cdot}7^5{\cdot}8^2)$ topology [11], indicating that the types of the dicarboxylate play important role in determining the structural diversity.

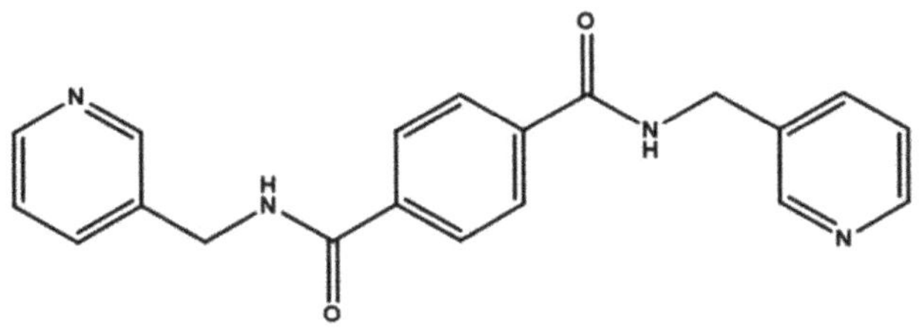

Figure 1. Structure of **L**.

The syntheses and crystal structures of [Co(**L**)(5-ter-IPA)(H_2O)$_2$]$_n$ (5-tert-H_2IPA = 5-tert-butylisophthalic acid), **1**, {[Co(**L**)(5-NO_2-IPA)]·2H_2O}$_n$ (5-NO_2-H_2IPA = 5-nitroisophthalic acid), **2**, {[Co(**L**)$_{0.5}$(5-NH_2-IPA)]·MeOH}$_n$ (5-NH_2-H_2IPA = 5-aminoisophthalic acid), **3**, {[Co(**L**)(MBA)]·2H_2O}$_n$ (H_2MBA = diphenylmethane-4,4′–dicarboxylic acid), **4**, {[Co(**L**)(SDA)]·H_2O}$_n$ (H_2SDA = 4,4-sulfonyldibenzoic acid), **5**, {[Co_2(**L**)$_2$(1,4-NDC)$_2$(H_2O)$_2$]·5H_2O}$_n$ (1,4-H_2NDC = naphthalene-1,4-dicarboxylic acid), **6**, {[Cd(**L**)(1,4-NDC)(H_2O)]·2H_2O}$_n$, **7**, and {[Zn_2(**L**)$_2$(1,4-NDC)$_2$]·2H_2O}$_n$, **8**, form the subject of this report. We observed that the roles of the dicarboxylate ligands and the metal atoms in the structural diversity of the CPs prepared thusly are significant. The governing factors of **1–3** in the degradation of methylene blue (MB) were also evaluated.

2. Results and Discussion

2.1. Synthesis

Complexes **1–8** were prepared by the hydro(solvo)thermal reactions of **L** with corresponding dicarboxylic acids and metal salts in different solvent systems at 100 °C for 48 h. Hydro(solvo)thermal synthesis enables a unique combination of pressure and temperature for crystallization of CPs. Characteristic FT-IR peaks for complexes **1–8** are N-H and C=O stretching which are from L. The range of N-H stretching is 3386–3483 cm^{-1}, probably coupled with the O-H stretching of the solvent molecule, while those around 1606–1653 cm^{-1} can be attributed to C=O stretching.

2.2. Crystal Structure of 1

The crystal structure of **1** conforms to the triclinic space group $P\bar{1}$ and the asymmetric unit consists of one Co(II) cation, two halves of an **L** ligand, one 5-ter-IPA^{2-} ligand, and two coordinated water molecules. The Co(II) metal center is coordinated by two pyridyl nitrogen atoms of two **L** ligands [Co-N = 2.1346(12) − 2.1816(12) Å], two oxygen atoms from two 5-ter-IPA^{2-} ligands [Co-O = 2.0587(10) and 2.1362(10) Å], and two coordinated water molecules [Co-O = 2.0794(10) and 2.1411(10) Å], forming a distorted octahedral geometry, as in Figure 2a. Two Co(II) ions are bridged by two 5-ter-IPA^{2-} ligands to form dinuclear units, which are connected by **L** ligands to afford a 2D layer. Considering the Co(II) cations as 4-coordinated nodes, 5-ter-IPA^2 as 2-connected nodes, and **L** ligands as linkers, the structure of **1** can be regarded as a 2,2,4-connected net with the point symbol $(12)(4{\cdot}12^5)(4)$ (standard representation), as in Figure 2b, determined using ToposPro [12]. Moreover, if the dinuclear units are considered as 3-coordinated nodes, the structure can be further simplified as a 3-connected net with the (6^3)-**hcb** topology (cluster representation) [13], as in Figure 2c.

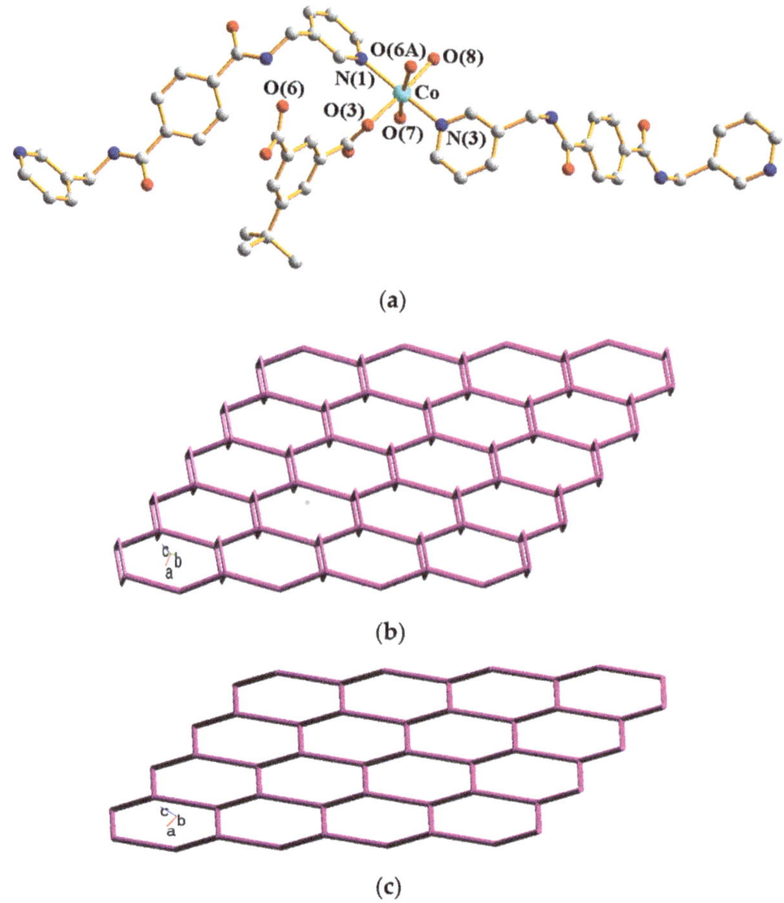

Figure 2. (**a**) A drawing showing the geometry of the Co(II) ion in **1**. Symmetry transformations: (A) −x + 1, −y + 1, −z. (**b**) A drawing showing the (12)($4{\cdot}12^5$)(4) topology. (**c**) A drawing showing the **hcb** topology.

2.3. *Crystal Structure of* 2

The structure of **2** was solved in the triclinic space group $P\bar{1}$ with one Co(II) cation, two halves of an **L** ligand, one 5-NO_2-IPA^{2-} ligand and two co-crystallized water molecules in each asymmetric unit. The Co(II) metal center is coordinated by four oxygen atoms from three 5-NO_2-IPA^{2-} ligands [Co-O = 2.007(2) – 2.240(2) Å] and two pyridyl nitrogen atoms from two **L** ligands [Co-N = 2.149(3) – 2.150(3) Å], resulting in a distorted octahedral geometry, as in Figure 3a. Two Co(II) ions are bridged by two 5-NO_2-IPA^{2-} ligands to form dinuclear units, which are connected by **L** ligands to afford a 3D framework. Considering the Co(II) cations as 5-coordinated nodes, 5-NO_2-IPA^{2-} as 3-coordinated nodes, and **L** ligands as linkers, the structure of **2** can be regarded as a 3,5-connected binodal 3D net with the point symbol of ($4^2{\cdot}6^5{\cdot}8^3$)($4^2{\cdot}6$)-3,5T1 (standard representation), as in Figure 3b. Moreover, if the dinuclear units are considered as 6-coordinated nodes, the structure can be further simplified as a 6-connected net with the ($4^{12}{\cdot}6^3$)-**pcu** topology (cluster representation), as in Figure 3c.

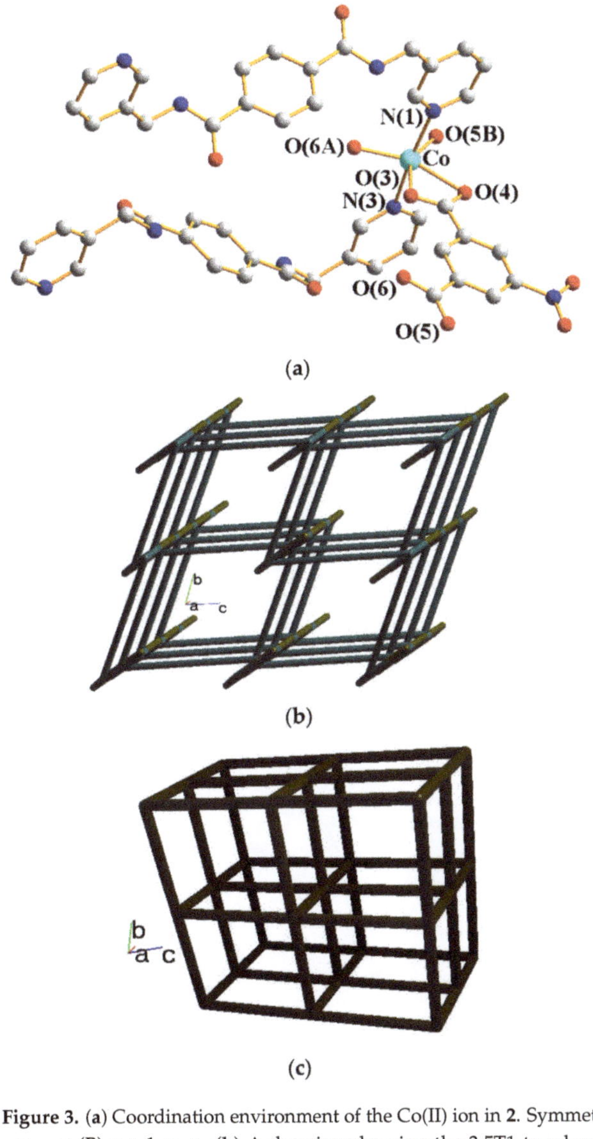

Figure 3. (**a**) Coordination environment of the Co(II) ion in **2**. Symmetry transformations: (A) $-x + 1, -y, -z$; (B) $x + 1, y, z$. (**b**) A drawing showing the 3,5T1 topology. (**c**) A drawing showing the **pcu** topology.

2.4. Crystal Structure of 3

Structural analysis demonstrates that **3** crystallizes in the triclinic $P\bar{1}$ space group. The asymmetric unit contains one Co(II) cation, half of an **L** ligand, one 5-NH_2-IPA^{2-}, and one co-crystallized MeOH molecule. The Co(II) metal center is coordinated by four oxygen atoms from three 5-NH_2-IPA $^{2-}$ ligands [Co-O = 2.0143(15) – 2.2104(14) Å], one pyridyl nitrogen atom from the **L** ligand, and one nitrogen from the 5-NH_2-IPA^{2-} ligand [Co-N = 2.1525(19) –2.2615(18) Å], showing a distorted octahedral geometry, as in Figure 4a. Two Co(II) ions are bridged by two 5-NH_2-IPA^{2-} ligands to form dinuclear units, which are connected by **L** ligands to afford a 2D layer. Considering the Co(II) cations as 4-connected nodes and 5-NH_2-IPA^{2-} ligands as 3-connected nodes, with **L** ligands as linkers, the structure of **3** can be simplified as a 3,4-connected 2D net with the $\{4^2{\cdot}6^3{\cdot}8\}\{4^2{\cdot}6\}$-**bey** topology (standard representation), as in Figure 4b. Moreover, if the dinuclear units are considered as 4-coordinated nodes, the structure can be further simplified as a 4-connected net with the $(4^4{\cdot}6^2)$-**sql** topology (cluster representation), as in Figure 4c.

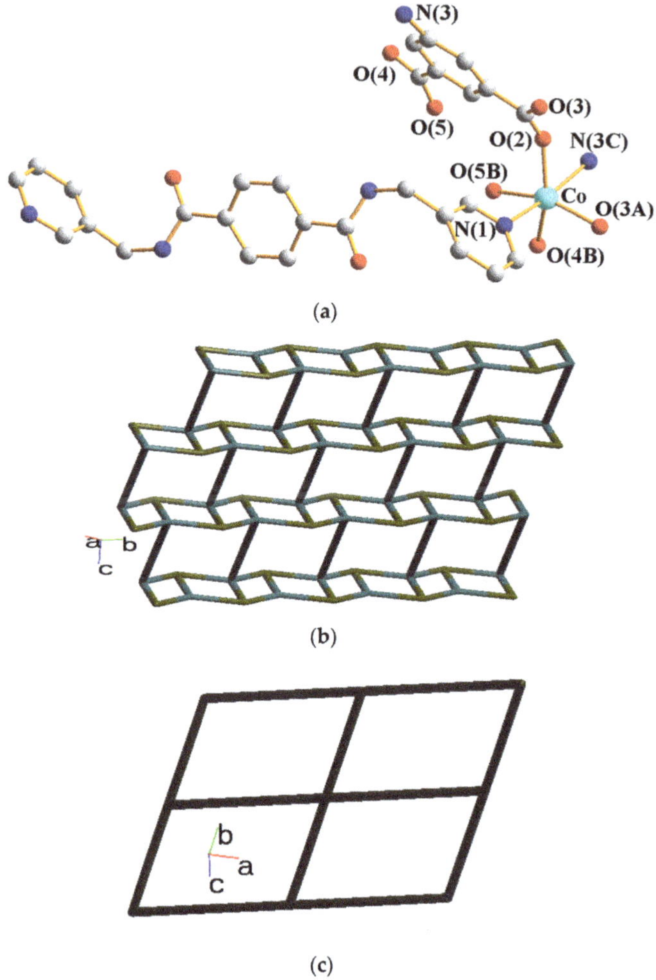

Figure 4. (**a**) Coordination environment about the Co(II) ion in **3**. Symmetry transformations: (A) −x − 1, −y, −z + 2; (B) −x − 1, −y + 1, −z + 2; (C) x − 1, y, z. (**b**) A drawing showing the **bey** topology. (**c**) A drawing showing the **sql** topology.

2.5. Crystal Structure of **4**

Single crystal X-ray diffraction of **4** conforms to the orthorhombic space group *Ibca*, and the asymmetric unit consists of half of a Co(II) ion, half of an **L** ligand, half of an MBA^{2-} ligand, and one co-crystallized water molecule. Figure 5a shows the coordination environment around the Co(II) metal center, which is six coordinated by two nitrogen atoms from two **L** ligands [Co-N = 2.083(3)] and four oxygen atom from two MBA^{2-} ligands [Co-O = 2.064(2)–2.263(3) Å], resulting in a distorted octahedral geometry. The Co(II) ions are interlinked by the **L** and MBA^{2-} ligands to give highly undulated 2D nets, as in Figure 5b. Topological analysis reveals that complex **4** forms 2-fold parallelly interpenetrated layers with the $\{4^4 \cdot 6^2\}$-**sql** topology, as in Figure 5c. In addition, layers of the 2-fold interpenetrated 2D layers polycatenated with other **sql** layers to form a final 2D → 3D entanglement, as in Figure 5d.

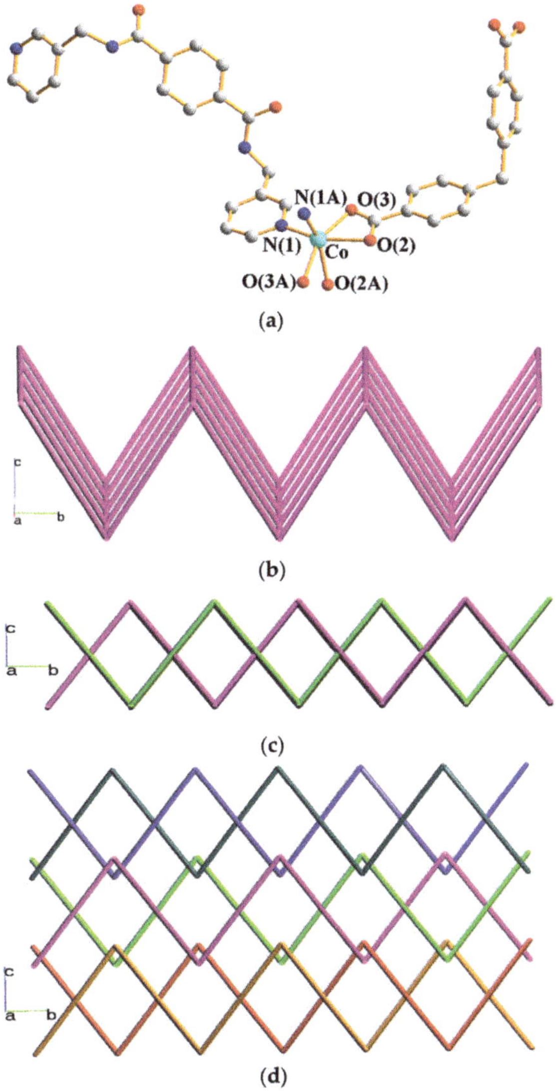

Figure 5. (**a**) Coordination environment around the Co(II) cation in **4**. Symmetry transformations used to generate equivalent atoms: (A) −x + 1, −y + 3/2, z. (**b**) A drawing showing the pleated 2D layer. (**c**) A drawing showing the 2-fold interpenetrated layers. (**d**) A drawing showing the polycatenation of 2-fold interpenetrated 2D nets.

2.6. Crystal Structures of 5

In the space group $P\bar{1}$, the structure of complex **5** was solved. The asymmetric unit consists of one Co(II) ion, half of an **L** ligand, one SDA^{2-} ligand, and one co-crystallized water molecule. Figure 6a shows the coordination environment around the dinuclear

Co(II) centers with a Co—Co distance of 2.8143(5). Both Co(1) and Co(2) are 5-coordinated by one pyridyl nitrogen atom of the **L** ligand [Co–N = 2.0564(17) Å] and four oxygen atoms of four SDA^{2-} ligands [Co–O = 2.0211(17) Å–2.0515(17) Å], resulting in a distorted square pyramidal geometry. Two Co(II) ions are bridged by four carboxylate groups of the SDA^{2-} ligands to form dinuclear paddlewheel units, which are further linked by the **L** ligands to form a 2D layer. If the dinuclear units are considered as 6-connected nodes, the SDA^{2-} ligands as 2-connected nodes, and the **L** ligands as linkers, the structure of **5** can be simplified as a 2D net with the $(4^2{\cdot}6^8{\cdot}8{\cdot}10^4)(4)_2$-2,6L1 topology, as in Figure 6b, which shows a 2-fold interpenetration, as in Figure 6c.

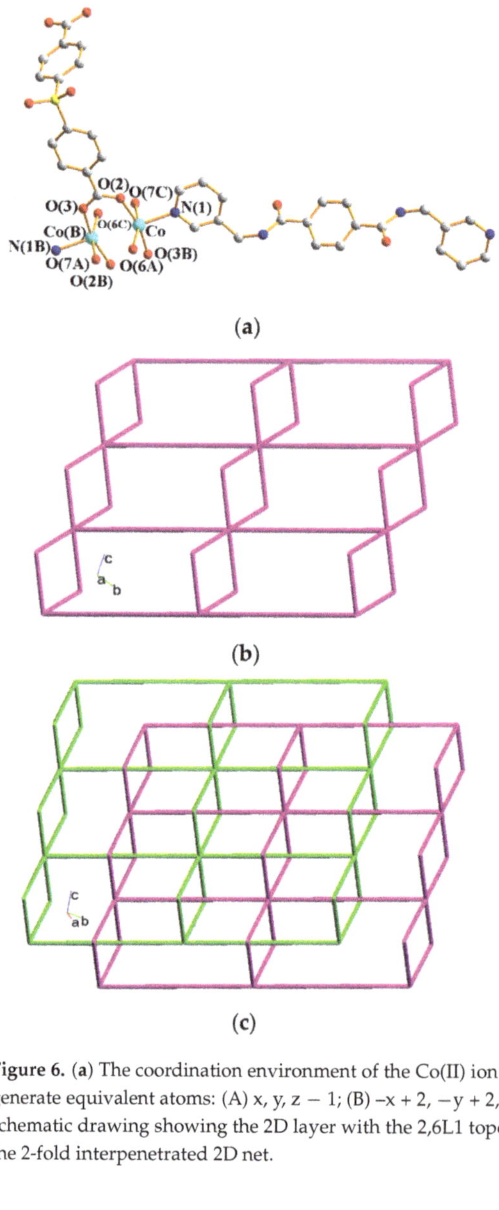

(c)

Figure 6. (**a**) The coordination environment of the Co(II) ion in **5**. Symmetry transformations used to generate equivalent atoms: (A) x, y, z − 1; (B) −x + 2, −y + 2, −z + 2; (C) −x + 2, −y + 2, −z + 3. (**b**) A schematic drawing showing the 2D layer with the 2,6L1 topology. (**c**) A schematic drawing showing the 2-fold interpenetrated 2D net.

2.7. Crystal Structures of 6

Crystals of **6** conform to the monoclinic space group $P2_1/c$ with each asymmetric unit consisting of one and two halves of a Co(II) cation, two **L** ligands, two 1,4-NDC^{2-} ligands, two coordinated water molecules, and five co-crystallized water molecules. Figure 7a shows the coordination environment of the Co(II) metal centers, which are all 6-coordinated. The Co(1) atom is coordinated by two nitrogen atoms from two **L** ligands [Co-N = 2.134(2) and 2.138(2) Å], three oxygen atoms from two 1,4-NDC^{2-} ligands [Co–O = 2.0257(18) Å – 2.19740515(19) Å], and one oxygen atom from the coordinated water molecule [Co-O = 2.0817(19) Å]. The Co(2) atom is located at the inversion center, which is coordinated by two pyridyl nitrogen atoms [Co–N = 2.114 (2) Å] from two **L** ligands and four oxygen atoms from two 1,4-NDC^{2-} ligands [Co–O = 2.1128(17)–2.1415(18) Å], whereas the Co(3) atom, which is also located at the inversion center, is coordinated by two nitrogen from two **L** ligands [Co–N = 2.145(2) Å], two oxygen atoms from two 1,4-NDC^{2-} ligands [Co–O = 2.0813(17)] and two oxygen atoms from two water molecules [Co–O = 2.1497(18)]. The Co(II) ions are linked by the 1,4-NDC^{2-} and **L** ligands to form a 3D framework. If the Co(II) ions are defined as 4-connected nodes and the **L** and 1,4-NDC^{2-} ligands as linkers, the structure of **6** can be simplified as a 4-connected net with the (6^5·8)-**cds** topology, as in Figure 7b.

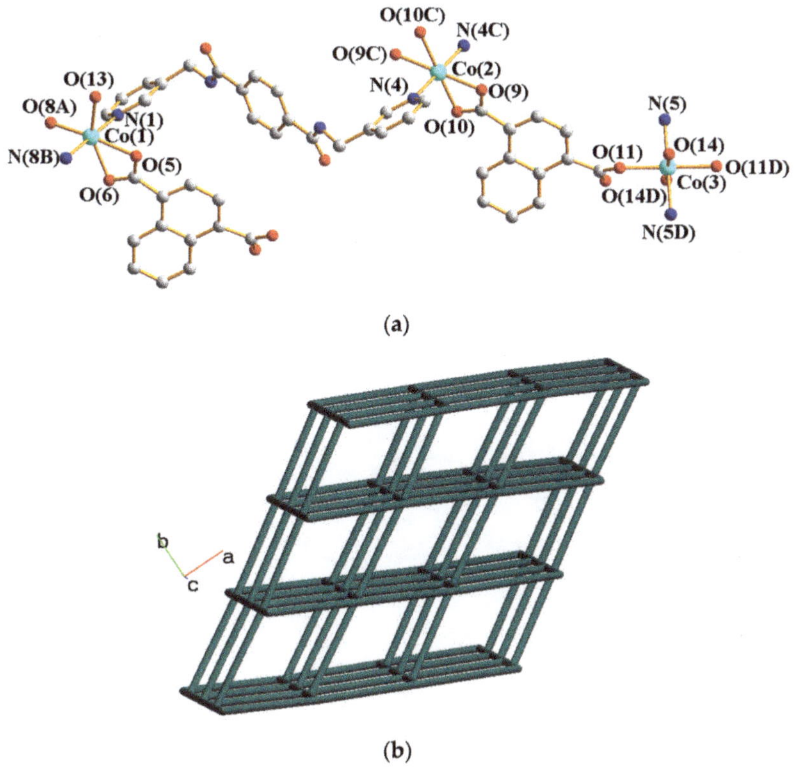

Figure 7. (**a**) The coordination environment of the Co(II) ion in **6**. Symmetry transformations used to generate equivalent atoms: (A) x, −y + 3/2, z + 1/2; (B) −x + 1, y − 1/2, −z − 1/2; (C) −x + 2, -y + 2, −z − 1; (D) −x + 2, −y + 2, −z − 2. (**b**) A drawing showing the **cds** topology.

2.8. Crystal Structures of **7**

Single-crystal X-ray diffraction analysis shows that **7** crystallizes in the monoclinic space group $P2_1/c$. The asymmetric unit is comprised of one Cd(II) cation, one **L** ligand, one 1,4-NDC^{2-} ligand, one coordinated water, and two lattice water molecules. Figure 8a depicts a drawing showing the coordination environment of the Cd(II) ion, which is 7-coordinated by four oxygen atoms from two 1,4-NDC^{2-} ligands, one oxygen atom from the water molecule [Cd–O = 2.343(2) − 2.396(19) Å], and two pyridyl nitrogen atoms from two **L** ligands [Cd–N = 2.319(2) and 2.410(3) Å]. The Cd (II) ions are further linked together by the **L** and 1,4-NDC^{2-} ligands to afford a 2D layer. If the Cd(II) cations are defined as 4-connected nodes and the 1,4-NDC^{2-} and **L** ligands are defined as 2-connected nodes, the structure of **7** can be simplified as a 2D net with the $\{4{\cdot}5^8\}\{4\}$-2,4L1 topology, as in Figure 8b.

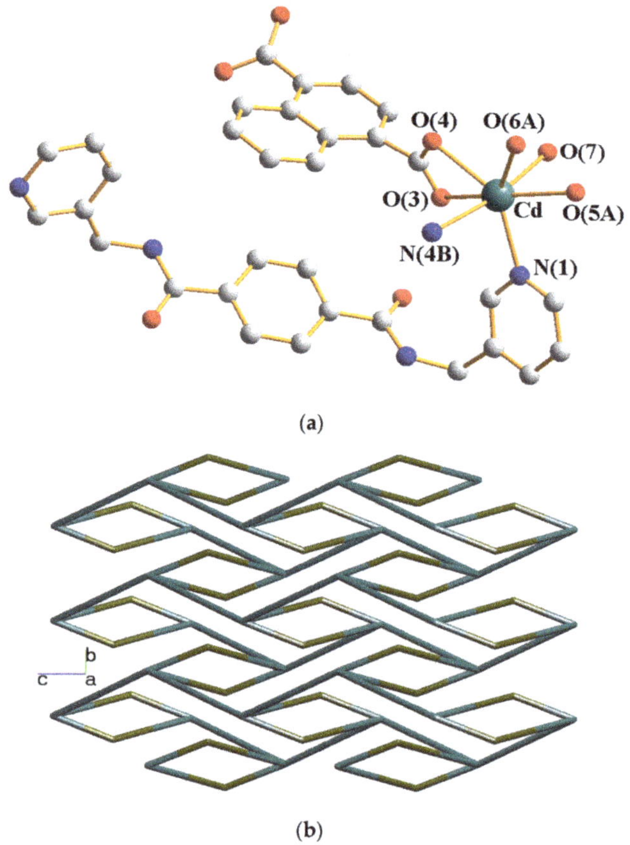

Figure 8. (**a**) The coordination environment of the Co(II) ion in **7**. Symmetry transformations used to generate equivalent atoms: (A) x, −y + 3/2, z + 1/2; (B) −x + 1, −y + 1, −z + 1. (**b**) A drawing showing the 2,4L1 topology.

2.9. Crystal Structures of **8**

The crystals of complex **8** conform to the triclinic space group $P\bar{1}$ with two Zn(II) ions, two **L** ligands, two 1,4-NDC^{2-} ligands, and two lattice water molecules in the asymmetric unit. Figure 9a shows the coordination environment of the Zn(II) centers. Both of the Zn(1) and Zn(2) atoms form tetrahedral geometries, which are 4-coordinated by two nitrogen atoms [Zn(1)–N(1) = 2.066(3); Zn(1)–N(4C) = 2.062(3) Å; Zn(1)–N(5) = 2.056(3) Å; Zn(1)–

N(8A) = 2.066(3) Å] from two **L** ligands and two oxygen atoms [Zn(1)–O(5) = 1.967(3) Å; Zn(1)–O(9) = 1.975(2); Zn(1)–O(7) = 1.969(2) Å; Zn(1)–O(12B) = 1.948(2)] from two 1,4-NDC^{2-} ligands. Topological analysis demonstrates that complex **8** displays a 2D net with the point symbol of $(10^2{\cdot}12)(10)_2(4{\cdot}10{\cdot}12^4)(4)$, as in Figure 9b.

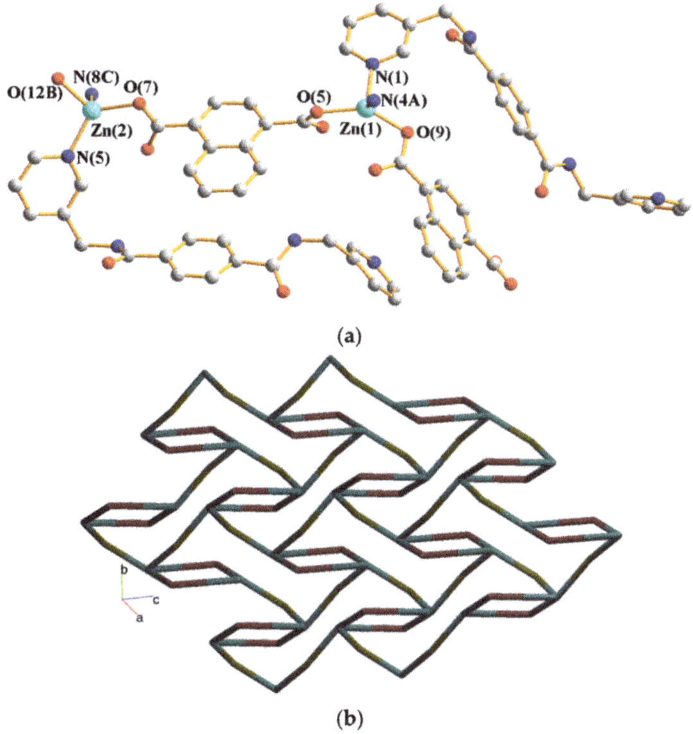

Figure 9. (**a**) The coordination environment of the Co(II) ion in **8**. Symmetry transformations used to generate equivalent atoms: (A) −x, −y + 2, −z + 2; (B) x, y, z − 1; (C) −x, −y + 1, −z + 1. (**b**) A drawing showing the $(10^2{\cdot}12)(10)_2(4{\cdot}10{\cdot}12^4)(4)$ topology.

2.10. Ligand Conformations and Coordination Modes

The **L** ligands in complexes **1–8** display various conformations which can be defined as follows: (A) the *cis* and *trans* conformations can be given if the two C=O groups are in the same and the opposite direction, respectively; (B) due to the different orientations adopted by the pyridyl nitrogen atoms and the amide oxygen atoms, three more conformations, namely *syn–syn*, *syn–anti*, and *anti–anti*, can also be found for bpba [8]. Table 1 lists the ligand conformations and coordination modes of the organic ligands in complexes **1–8**. The **L** ligands in **1–8** bridge two metal ions through two pyridyl nitrogen atoms, adopting five different conformations including *trans anti–anti*, *cis syn–syn*, *trans syn–syn*, *trans syn–anti* and *cis anti–anti*. On the other hand, the dicarboxylate ligands in **1–8** bridge two to four metal ions with various coordination modes.

Table 1. Ligand conformations and bonding modes of **1–8**.

	Conformation	Coordination Mode
1	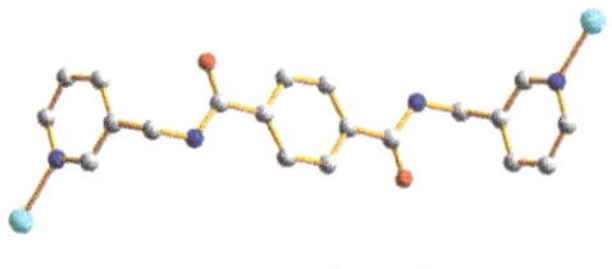 *trans anti–anti* 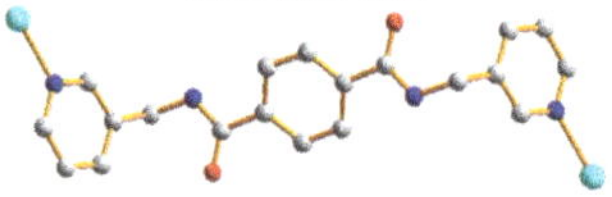 *trans anti–anti*	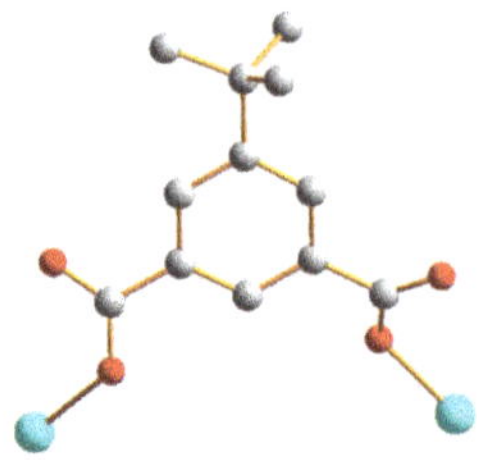 μ_2-κO:κO′
2	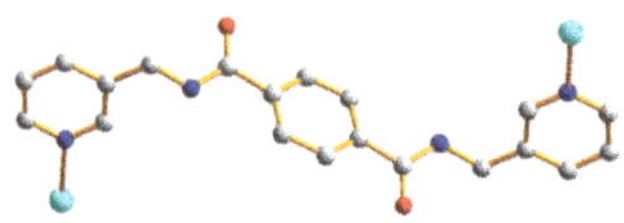 *trans anti–anti*  *trans syn–syn*	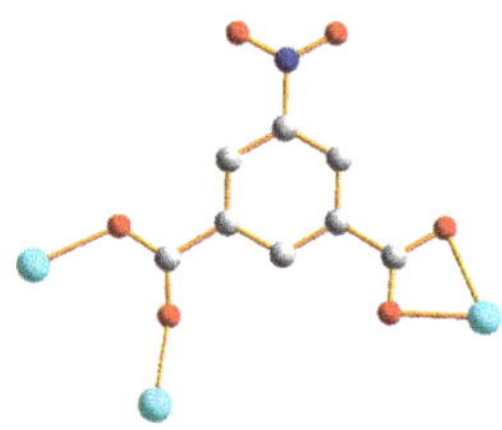 μ_3-κ^{2}O,O′:κO″:κO‴
3	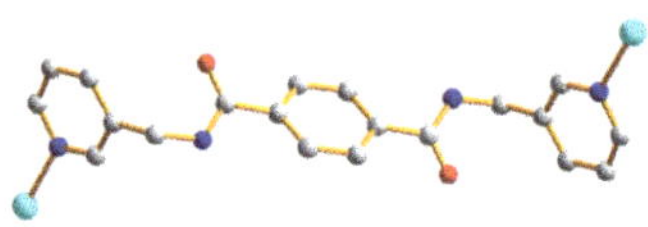 *trans anti–anti*	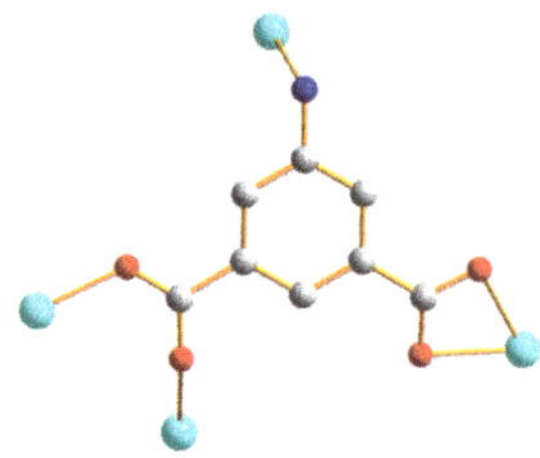 μ_4-κ^{2}O,O′:κO″:κO‴:κN
4	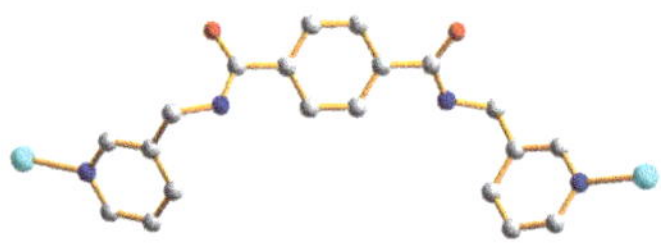 *cis syn–syn*	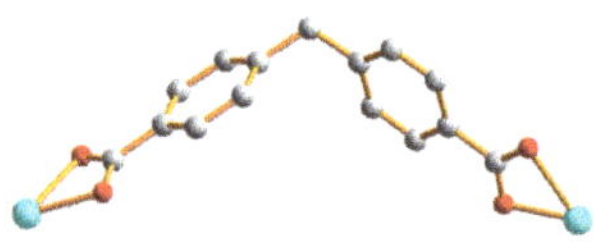 μ_2-κ^{2}O,O′: κ^{2}O″,O‴
5	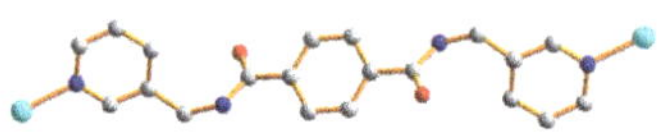 *trans anti–anti*	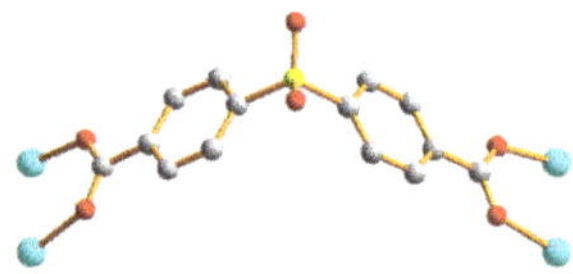 μ_4-κO:κO′:κO″:κO‴

Table 1. *Cont.*

	Conformation	Coordination Mode
6	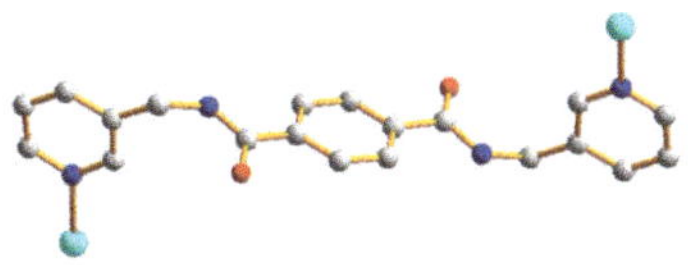 *trans syn–syn* 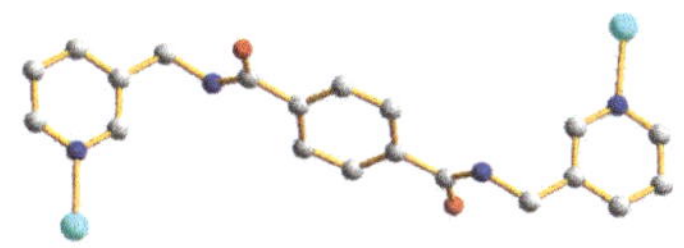 *trans anti–anti*	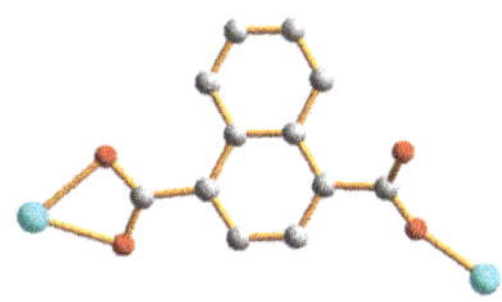 μ_2-κ^2O,O':$\kappa O''$
7	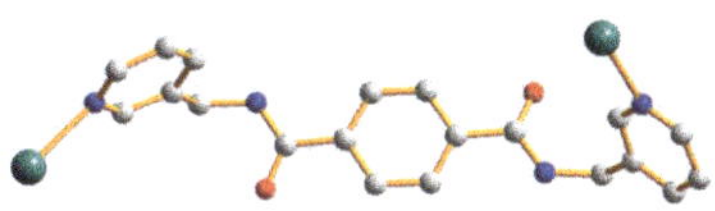 *trans syn–syn*	 μ_2-κ^2O,O': κ^2O'',O'''
8	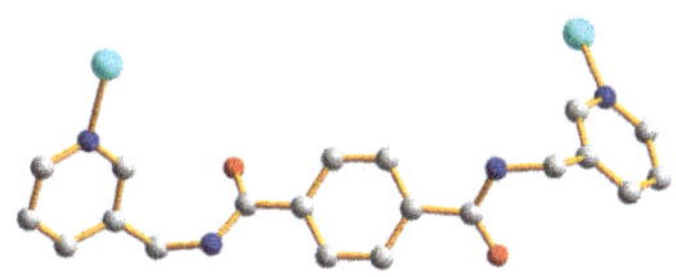 *trans syn–anti* 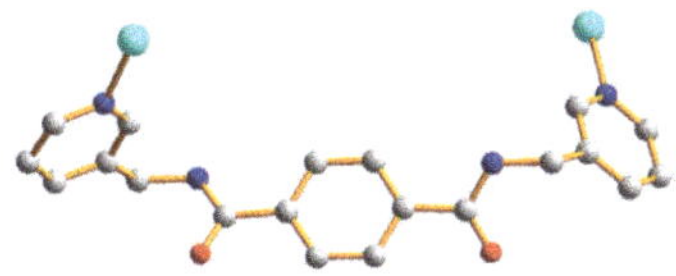 *cis anti–anti*	 μ_2-κO:$\kappa O'$

2.11. Structural Comparisons

Structural comparisons of complexes **1–8** show that the structural diversity is subject to the change in the dicarboxylate ligand. The different structural types in **1–3** demonstrate the substituent effect of the group at the fifth position of the phenyl ring. The use of the angular dicarboxylic acids, such as H_2MBA and H_2SDA, give entangled **4** and **5**, showing a polycatenated net of 2-fold interpenetration and a 2-fold interpenetrated net, respectively. The metal effect on the structural diversity is shown in **6–8** by changing the metal atom from Co, Cd, to Zn, giving **cds**, 2,4L1, and $(10^2{\cdot}12)(10)_2(4{\cdot}10{\cdot}12^4)(4)$ topologies, respectively.

2.12. Photodegradation

The governing role of CPs in the photodegradation of organic pollutants has been a subject of current interest [14–19]. Complexes **1–3**, which differ in the fifth position of the phenyl ring of the dicarboxylate ligands, *i.e.*, the *tert*-butyl, NO_2 and NH_2 groups, respectively, thus, provide a unique opportunity to compare the substituent effect on the photodegradation. Methylene blue (MB, $C_{16}H_{18}ClN_3S$) was selected as the dye contaminant, and the experiments were carried out with 30 wt % H_2O_2 under 365 nm UV light. Time-dependent absorption spectra of the MB solutions under 365 nm UV light are provided as Figures S1–S8.

The intensity of the peculiar absorption band at 663 nm was utilized to precisely monitor the degradation process of MB. Figure 10 illustrates the variations in the A_t/A_0 of MB solutions vs irradiation time for complexes **1–3**, showing that the absorption intensities of MB reduced gradually with increasing reaction time, where A_0 is the initial absorbance of the MB solution and A_t is the absorbance of the solution after illumination at time t. Degradation efficiency (DE) of MB was calculated by using DE % = $[(A_0 - A_t)/A_0] \times 100$. Additionally, the DE % with the mean values and standard deviations were evaluated, Tables S1–S8. After 120 min, the DE of MB for the various strategies are as follows: 3% (blank), 53.46% (MB + H_2O_2), 6% (MB + complex **1**), 10.42% (MB + complex **2**), 16.43% (MB + complex **3**), 61.35% (MB + H_2O_2 + complex **1**), 77.59% (MB + H_2O_2 + complex **2**), and 95.06% (MB + H_2O_2 + complex **3**), demonstrating that the DE % of MB by the complexes participated with H_2O_2 follows the pattern of **1** < **2** < **3**. Moreover, the Brunauer–Emmett–Teller (BET) surface areas obtained from the N_2 adsorption experiments were 4.96, 6.12, and 9.95 m^2/g for **1–3**, respectively, as in Figures S9–S11. The PXRD patterns of complexes **1–3** succeeding photodegradation processes were examined. No noticeable alterations were found for **1** and **2**, whereas significant change has been observed for **3**, as illustrated in Figures S12–S14. The structural change in **3** may enhance the photodegradation efficiency. Structural modification was also observed for **3** after the N_2 adsorption and desorption, as in Figures S15–S17, indicating that complex **3** was not stable during the experiments.

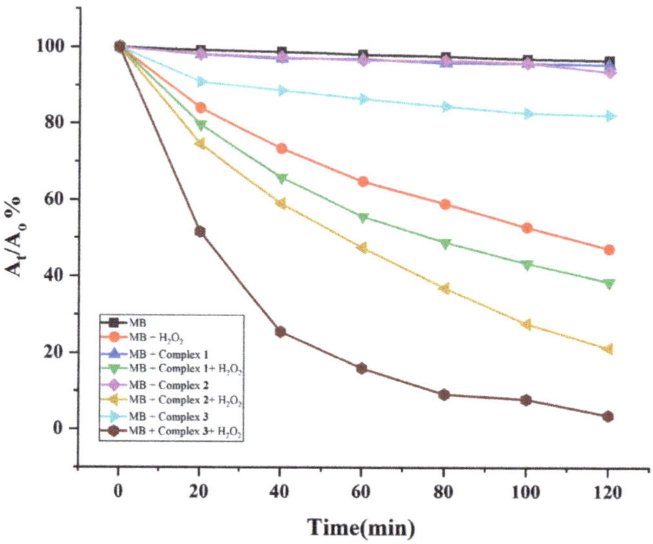

Figure 10. Photodegradation rates of MB solution under UV irradiation.

Although the role of the *tert*-butyl, NO_2, and NH_2 groups in determining the DE is complicated, the different BET surface areas of the original **1–3** resulting from the different substituent groups can be influential. The hydroxyl radical (OH·) has been considered as the major oxidant which decomposes the organic dye with a good efficiency [15]. High surface area reflects a higher adsorption quantity of H_2O_2 that led to the formation of (OH·) and, thus, implies more MB can be degraded. For comparisons it is noted that the CPs $\{[Zn(L2)(AIPA)]\cdot 2H_2O\}_n$ (L2 = *N,N′*-bis(3-pyridinyl)terephthalamide; H_2AIPA = 5-acetamidoisophthalic acid) and $\{[Zn(L3)(AIPA)]\cdot 2H_2O\}_n$ (L3 = *N,N′*-di(3-pyridyl)adipoamide), which adopted self-catenated 3D frameworks with the $(4^{24}\cdot 6^4)$-8T2 and the $(4^4\cdot 6^{10}\cdot 8)$-**mab** topologies, respectively, promoted the MB degradation, and the DE were 81.56 and 85.46%, respectively [20]. On the other hand, the four topologically identical CPs having the 2-fold interpenetrating 3D net with the **mog** topology,

{[M(L4)$_{0.5}$(L5)(H_2O)$_2$]·H_2O}$_n$ (M = Co and Ni; H_4L4 = bis(3,5-dicarboxyphenyl)adipoamide; L5 = bis(N-pyrid-3-ylmethyl) adipoamide) and {[M_2(L4)(L6)$_2$(H_2O)$_4$]·3H_2O}$_n$ (M = Co and Ni; L6 = bis(N-pyrid-3-ylmethyl) suberoamide) also display good photodegradation performance toward MB, and the Co(II) CPs display better catalytic ability than the Ni(II) ones [21].

3. Materials and Methods

3.1. General Procedures

Elemental analyses involving C, H, and N atoms were performed on a PE 2400 series II CHNS/O (PerkinElmer instruments, Shelton, CT, USA) or an Elementar Vario EL-III analyzer (Elementar Analysensysteme GmbH, Hanau, Germany). Infrared spectra were obtained from a JASCO FT/IR-460 plus spectrometer with pressed KBr pellets (JASCO, Easton, MD, USA). Powder X-ray diffraction patterns were carried out with a Bruker D8-Focus Bragg–Brentano X-ray powder diffractometer equipped with a CuKα (λ_α = 1.54178 Å) sealed tube (Bruker Corporation, Karlsruhe, Germany).

3.2. Materials

The reagents Co(OAc)$_2$·4H_2O, Cd(OAc)$_2$·H_2O and Zn(OAc)$_2$·2H_2O were purchased from Alfa Aesar (Ward Hill, MA, USA), whereas 5-tert-butylisophthalic acid (5-tert-H_2IPA), 5-nitroisophthalic acid (5-NO_2-H_2IPA), 5-aminoisophthalic acid (5-NH_2-H_2IPA), diphenylmethane-4,4′–dicarboxylic acid (H_2MBA), 4,4-sulfonyldibenzoic acid (H_2SDA), and naphthalene-1,4-dicarboxylic acid (1,4-H_2NDC) were from Aldrich Chemical Co. (St. Louis, MO, USA). The ligand *N,N′*-bis(pyridin-3-ylmethyl)terephthalamide (**L**) was prepared according to a published procedure [22].

3.3. Preparations

3.3.1. [Co(**L**)(5-ter-IPA)(H_2O)$_2$]$_n$, **1**

A 23 mL Teflon-lined steel autoclave was sealed with Co(OAc)$_2$·4H_2O (0.050 g, 0.20 mmol), **L** (0.070 g, 0.20 mmol), 5-tert-H_2IPA (0.042 g, 0.20 mmol), and 10mL H_2O, which was heated to 100 °C for two days and then cooled to room T at a rate of 2 °C per hour. Orange crystals formed, which were collected and purified. Yield: 0.061 g (46%). Anal. calcd for $C_{32}H_{34}CoN_4O_8$ (MW = 661.56): C, 58.1; H, 5.1; N, 8.5%. Found: C, 58.8; H, 5.3; N, 8.7%. FT-IR (cm^{-1}): 3420(s), 2965(w), 1663(m), 1606(m), 1536(m), 1479(m), 1431(m), 1372(m), 1281(w), 1187(w), 1107(m), 1038(w), 933(m), 786(s), and 710(s).

3.3.2. {[Co(**L**)(5-NO_2-IPA)]·2H_2O}$_n$, **2**

Purple crystals of **2** were prepared by following similar procedures for **1**, except that 5-NO_2-H_2IPA (0.042 g, 0.20 mmol) was used. Yield: 0.065 g (50%). Anal. calcd for $C_{28}H_{25}CoN_5O_{10}$ (MW = 650.46): C, 51.7; H, 3.9; N, 10.8%. Found: C, 52.2; H, 3.6; N, 10.8%. FT-IR (cm^{-1}) :3473(s),3386(s),3263(s),3083(m), 1647(m), 1613(m), 1556(m), 1530(m), 1464(m), 1392(m), 1346(m), 1288 (w), 1083(w), 1038(w), 996(m), 736(s), 718(s), and 702(s).

3.3.3. {[Co(**L**)$_{0.5}$(5-NH_2-IPA)]·MeOH}$_n$, **3**

Complex **3** was prepared by following similar procedures for **1**, except that 5-NH_2-H_2IPA (0.036 g, 0.20 mmol) in 10 mL of MeOH/H_2O was used. Red crystals were collected. Yield: 0.064 g (72%). Anal. calcd for $C_{19}H_{18}CoN_3O_6$ (MW = 443.29): C, 51.5; H, 4.1; N, 9.5%. Found: C, 51.2; H, 3.7; N, 9.5%. FT-IR (cm^{-1}): 3389(s),3326(s), 1653(m), 1546(m), 1530(m), 1451(m), 1404(m), 1346(m), 1278 (w), 1055(w), 1033(w), 957(m), 781(s), 729 (s), and 711(s).

3.3.4. {[Co(**L**)(MBA)]·2H_2O} $_n$, **4**

Complex **4** was prepared by following similar procedures for **1**, except that a mixture of Co(OAc)$_2$·4H_2O (0.05 g, 0.20 mmol), **L** (0.070 g, 0.20 mmol), and MBA (0.052 g, 0.20 mmol) in 10 mL of H_2O was used. Purple crystals were obtained. Yield: 0.084 g (60%). Anal. calcd for $C_{35}H_{32}CoN_4O_8$ (MW = 695.57): C, 60.4; H, 4.6; N, 8.1%. Found: C, 60.7; H, 4.3; N,

7.7%. FT-IR (cm^{-1}): 3450 (s), 2920(m),2850(m), 1629(m), 1613(m), 1469(s), 1392(m), 1358(m), 1301(w), 871(m), 760(s), 727(w), and 702(w).

3.3.5. {[Co(**L**)(SDA)]·H_2O}$_n$, **5**

Complex **5** was prepared by following similar procedures for **1**, except that a mixture of $Co(OAc)_2.4H_2O$ (0.025 g, 0.10 mmol), **L** (0.035 g, 0.10 mmol), SDA (0.031 g, 0.10 mmol), and 8 mL of H_2O in 2 mL of MeOH was used. Violet crystals were obtained. Yield: 0.023 g (41%). Anal. calcd for $C_{24}H_{19}CoN_2O_8S$ (MW = 554.40): C, 52.0; H, 3.4; N, 5.1%. Found: C, 51.9; H, 3.4; N, 6.5%. The large inconsistency of the N atom may be due to the fact that the crystals used for measurement suffered the loss of co-crystallized solvents or the contamination of minor product which was not able to be removed. FT-IR (cm^{-1}): 3428(s), 2925(s), 2851(m), 3083(m), 1648(m), 1549(m), 1418(m), 1301(m), 1173(m), 1127(m), 1035(w), 1288 (w), 992(w), 862(w), 762 (s), and 700(s).

3.3.6. {[Co_2(**L**)$_2$(1,4-NDC)$_2$(H_2O)$_2$]·5H_2O}$_n$, **6**

Complex **6** was prepared by following similar procedures for **1**, except that a mixture of $Co(OAc)_2·4H_2O$ (0.025 g, 0.10 mmol), 1,4-H_2NDC (0.022 g, 0.10 mmol), and **L** (0.035 g, 0.10 mmol) in 10 mL H_2O was used, and the reaction was carried out at 80 °C. Pink crystals were obtained. Yield: 0.055 g (81%). Anal. calcd for $C_{64}H_{62}Co_2N_8O_{19}$ (MW = 1365.07): C, 56.3; H, 4.6; N, 8.2%. Found: C, 56.8; H, 4.3; N, 8.8%. IR (cm^{-1}): 3483(s), 3314(s), 2915(m), 1649(s), 1546(s), 1428(m), 1349(m), 1295(m), 1256(w), 1198(w), 1051(m), 977(s), and 838(m).

3.3.7. {[Cd(**L**)(1,4-NDC)(H_2O)]·2H_2O}$_n$, **7**

Prepared as described for **6**, except that $Cd(OAc)_2·H_2O$ (0.027 g, 0.10 mmol) was used. Colorless crystals were obtained. Yield: 0.039 g (54%). Anal. calcd for $C_{32}H_{30}CdN_4O_9$ (MW = 727.00): C, 52.7; H, 4.1; N, 7.7%. Found: C, 52.7; H, 3.8; N, 7.5%. IR (cm^{-1}): 3436(s), 1643(s), 1559(m), 1428(s), 1367(s), 1284(m), 1234(m), 1189(m), 1117(w), 1058(w), 873(m), 846(s), and 700(m).

3.3.8. {[Zn_2(**L**)$_2$(1,4-NDC)$_2$]·2H_2O}$_n$, **8**

Complex **8** was prepared by following similar procedures to those for **6**, except that a mixture of $Zn(OAc)_2·2H_2O$ (0.022g, 0.10 mmol) was used. Colorless crystals were obtained. Yield: 0.061 g (95%). Anal. calcd for $C_{64}H_{52}Zn_2N_8O_{14}$ (MW = 1287.87): C, 59.7; H, 4.1; N, 8.7%. Found: C, 60.4; H, 3.7; N, 8.7%. IR (cm^{-1}): 3330(s), 3060(m), 2929(w), 1644(s), 1600(s), 1540(s), 1430(m), 1328(m), 1291(w), 1257(w), 1193(m), 1123(w), 833(m), 786(m), and 701(m).

3.4. *Powder X-ray Analysis and IR Spectra*

In order to check the phase purity of the product, powder X-ray diffraction (PXRD) experiments were carried out for complexes **1–8**. As shown in Figures S18–S25, the peak positions of the experimental and simulated PXRD patterns were in a good agreement with each other, indicating their bulk purities. The IR spectra of complexes **1–8** are provided in the Supplementary Materials as Figure S26.

3.5. *Procedures for Photodegradation*

The experiments were carried out in a homemade photodegradation box (Figure S27). For the experiments, test tube 1 (blank), tube 2 (0.1 mL H_2O_2), tube 3 (10 mg complex), and tube 4 (10 mg complex + 0.1 mL H_2O_2) were prepared. A total of 10 mL of a 10 ppm MB solution was added to each tube, which was prepared by diluting 10 mg MB with deionized water in a 1000 mL quantitative bottle. Each tube was then irradiated with the 365 nm UV light for 20, 40, 60, 80, 100, and 120 min, respectively, and then their absorption spectra were measured. Tube 3 and tube 4 were first stirred in the dark for 15 min to confirm the physical adsorption of the complex.

3.6. X-ray Crystallography

Single-crystal X-ray diffraction data for complexes **1–8** were collected on a Bruker AXS SMART APEX II CCD diffractometer with graphite-monochromated MoKα (λ_α = 0.71073 Å) radiation at 296 K [23]. Data reduction and absorption correction were performed by using standard methods with well-established computational procedures. Some of the heavier atoms were located by the direct or Patterson method, and the remaining atoms were found in a series of Fourier maps and least-squares refinements, while the hydrogen atoms were added by using the HADD command in SHELXTL [24]. Table 2 lists the basic information pertaining to crystal parameters and structure refinement. CCDC no. 2238099–2238106 contain the supplementary crystallographic data for this paper. These data can be obtained free of charge via http://www.ccdc.cam.ac.uk/conts/retrieving.html or from the Cambridge Crystallographic Data Centre, 12 Union Road, Cambridge CB2 1EZ, UK; fax: +44 1223 336 033; e-mail: deposit@ccdc.cam.ac.uk; or at: http://www.ccdc.cam.ac.uk.

Table 2. Crystallographic data for **1–8**.

Compound	**1**	**2**	**3**
Formula	$C_{32}H_{34}N_4O_8Co$	$C_{28}H_{25}N_5O_{10}Co$	$C_{19}H_{18}CoN_3O_6$
Formula weight	661.56	650.46	443.29
Crystal system	Triclinic	Triclinic	Triclinic
Space group	$P\bar{1}$	$P\bar{1}$	$P\bar{1}$
a, Å	10.3115(3)	10.0958(2)	8.3283(3)
b, Å	11.9684(4)	11.3920(3)	10.0484(3)
c, Å	13.2748(4)	12.6159(3)	11.3130(4)
α, °	74.6853(9)	73.5612(11)	81.041(2)
β, °	73.1745(9)	84.2220(10)	87.225(25)
γ, °	73.1501(9)	87.1266(11)	82.0898(19)
V, Å^3	1471.49(8)	1384.23(6)	925.94(5)
Z	2	2	2
D_{calc}, Mg/m^3	1.493	1.561	1.590
F (000)	690	670	456
μ (Mo K_α), mm^{-1}	0.644	0.689	0.971
Range (2θ) for data collection, deg	$3.270 \leq 2\theta \leq 56.854$	$3.380 \leq 2\theta \leq 56.662$	$3.646 \leq 2\theta \leq 56.688$
Independent reflections	7357 [R(int) = 0.0247]	6880 [R(int) = 0.0226]	4607 [R(int) = 0.0329]
Data/restraints/parameters	7357/0/406	6880/0/397	4607/0/262
Quality-of-fit indicator [c]	1.035	1.065	1.005
Final R indices [I > 2σ(I)] [a,b]	R1 = 0.0303 wR2 = 0.0732	R1 = 0.0536 wR2 = 0.1556	R1 = 0.0378 wR2 = 0.0824
R indices (all data)	R1 = 0.0369, wR2 = 0.0762	R1 = 0.0656, wR2 = 0.1659	R1 = 0.0548, wR2 = 0.0895
Compound	**4**	**5**	**6**
Formula	$C_{35}H_{32}CoN_4O_8$	$C_{24}H_{19}CoN_2O_8S$	$C_{64}H_{62}Co_2N_8O_{19}$
Formula weight	695.57	554.40	1365.07
Crystal system	Orthorhombic	Triclinic	Monoclinic

Table 2. *Cont.*

Space group	*Ibca*	$P\bar{1}$	$P2_1/c$
a, Å	13.6153(3)	9.1565(4)	16.5329(3)
b, Å	19.3800(5)	10.9657(5)	17.1867(3)
c, Å	25.1236(6)	13.2184(6)	21.3467(4)
α, °	90	97.8126(15)	90
β, °	90	109.4504(17)	103.3525(9)
γ, °	90	104.9062(15).	90
V, Å^3	6629.2(3)	1173.06(9)	5901.61(19)
Z	8	2	4
D_{calc}, Mg/m^3	1.394	1.570	1.536
F (000)	2888	568	2832
μ (Mo K_α), mm^{-1}	0.576	0.874	0.649
Range (2θ) for data collection, deg	$3.242 \leq 2\theta \leq 52.000$	$3.962 \leq 2\theta \leq 56.714$	$3.076 \leq 2\theta \leq 56.588$
Independent reflections	3267 [R(int) = 0.0767]	5836 [R(int) = 0.0234]	14613 [R(int) = 0.0466]
Data/restraints/parameters	3267/1/223	5836/0/334	14613/0/841
Quality-of-fit indicator [c]	1.014	1.067	1.060
Final R indices [I > 2σ(I)] [a,b]	R1 = 0.0492, wR2 = 0.1109	R1 = 0.0375, wR2 = 0.1079	R1 = 0.0526, wR2 = 0.1264
R indices (all data)	R1 = 0.1309, wR2 = 0.1409	R1 = 0.0493, wR2 = 0.1155	R1 = 0.0892, wR2 = 0.1448
Compound	**7**	**8**	
Formula	$C_{32}H_{30}CdN_4O_9$	$C_{64}H_{52}Zn_2N_8O_{14}$	
Formula weight	727.00	1287.87	
Crystal system	Monoclinic	Triclinic	
Space group	$P2_1/c$	$P\bar{1}$	
a, Å	16.9732(3)	a = 12.8606(12)	
b, Å	9.4163(2)	b = 13.1507(14)	
c, Å	20.7184(4)	c = 18.7726(18)	
α, °	90	90.177(6)	
β, °	114.0812(10)	101.414(6)	
γ, °	90	113.475(5)	
V, Å^3	3023.12(10)	2842.7(5)	
Z	4	2	
D_{calc}, Mg/m^3	1.597	1.505	
F(000)	1480	1328	
μ(Mo K_α), mm^{-1}	0.786	0.922	
Range (2θ) for data collection, deg	$2.628 \leq 2\theta \leq 56.620$	$2.222 \leq 2\theta \leq 56.840$	
Independent reflections	7521 [R(int) = 0.0571]	14194 [R(int) = 0.0806]	
Data/restraints/parameters	7521/0/419	14194/0/793	

Table 2. *Cont.*

Quality-of-fit indicator [c]	1.009	1.004
Final R indices $[I > 2\sigma(I)]$ [a,b]	R1 = 0.0373, wR2 = 0.0670	R1 = 0.0567, wR2 = 0.1012
R indices (all data)	R1 = 0.0672, wR2 = 0.0763	R1 = 0.1485, wR2 = 0.1265

[a] $R_1 = F_o - F_c/F_o$. [b] $wR_2 = [w(F_o^2 - F_c^2)^2/w(F_o^2)^2]^{1/2}$. $w = 1/[\sigma^2(F_o^2) + (ap)^2 + (bp)]$, $p = [max(F_o^2 \text{ or } 0) + 2(F_c^2)]/3$. a = 0.0329, b = 0.9214, **1**; a = 0.0873, b = 1.5032, **2**; a = 0.0413, b = 0.5057, **3**; a = 0.0644, b = 0.1557, **4**; a = 0.0665, b = 0.5236, **5**; a = 0.0590, b = 6.0526, **6**; a = 0.0259, b = 1.0874, **7**; a = 0.0475, b = 0.1495, **8**. [c] quality-of-fit = $[\Sigma w(|F_o^2| - |F_c^2|)^2]/(N_{observed} - N_{parameters})^{1/2}$.

4. Conclusions

Eight divalent CPs constructed from **L** and various dicarboxylic acids have been successfully accomplished. The changes in the substituted group at the fifth position of the phenyl rings of the dicarboxylic acids from *tert*-butyl and NO_2 to the NH_2 group drastically alters the structural types, affording the simplified structures with the **hcb**, **pcu**, and **sql** topologies for complexes **1–3**, respectively. The use of the angular dicarboxylic acids, such as H_2MBA and H_2SDA, gave entangled CPs **4** and **5**, showing a polycatenation of a 2-fold interpenetrated 2D layer with the **sql** and a 2-fold interpenetrated 2D layer with the 2,6L1 topologies, whereas the metal effect on the structural diversity can be shown in complexes **6–8** by changing the metal atom from Co, Cd to Zn, affording a 3D framework with the **cds**, a 2D layer with the 2,4L1, and a 2D layer with the $(10^2{\cdot}12)(10)_2(4{\cdot}10{\cdot}12^4)(4)$ topologies, respectively. The structural diversity of the semi-rigid **L**-based divalent CPs is, thus, subject to the identities of the metal atom and the dicarboxylic acid. The degradation efficiency toward MB that follows **1** < **2** < **3** can be ascribed to their increasing surface areas, resulting from the different substituent groups of *tert*-butyl, NO_2, and NH_2 at the fifth position of the phenyl ring of the respective dicarboxylate ligands.

Supplementary Materials: The following supporting information can be downloaded at: https://www.mdpi.com/article/10.3390/molecules28052226/s1, UV-vis spectra (Figures S1–S8). Degradation efficiency (Tables S1–S8). N_2 sorption isotherm (Figures S9–S11). PXRD patterns (Figures S12–S25). IR spectra (Figure S26). Photodegradation box (Figure S27).

Author Contributions: Investigation, M.G.; data curation, S.-Y.Z.; review and supervision, C.-H.L. and J.-D.C. All authors have read and agreed to the published version of the manuscript.

Funding: This research was funded by the National Science and Technology Council of the Republic of China: NSC 109-2113-M-033-009.

Institutional Review Board Statement: Not applicable.

Informed Consent Statement: Not applicable.

Data Availability Statement: Data are contained within the article or Supplementary Materials.

Acknowledgments: We are grateful to the National Science and Technology Council of the Republic of China for support.

Conflicts of Interest: The authors declare no conflict of interest.

Sample Availability: Samples of the compounds are available from the authors.

References

1. Lustig, W.P.; Mukherjee, S.; Rudd, N.D.; Desai, A.V.; Li, J.; Ghosh, S.K. Metal-organic frameworks: Functional luminescent and photonic materials for sensing applications. *Chem. Soc. Rev.* **2017**, *46*, 3242–3285. [CrossRef]
2. Li, B.; Wen, H.-M.; Cui, Y.; Zhou, W.; Qian, G.; Chen, B. Emerging Multifunctional Metal-Organic Framework Materials. *Adv. Mater.* **2016**, *28*, 8819–8860. [CrossRef] [PubMed]
3. Wales, D.J.; Grand, J.; Ting, V.P.; Burke, R.D.; Edler, K.J.; Bowen, C.R.; Mintova, S.; Burrows, A.D. Gas sensing using porous materials for automotive applications. *Chem. Soc. Rev.* **2015**, *44*, 4290–4321. [CrossRef] [PubMed]
4. Kreno, L.E.; Leong, K.; Farha, O.K.; Allendorf, M.; Van Duyne, R.P.; Hupp, J.T. Metal–organic framework materials as chemical sensors. *Chem. Rev.* **2012**, *112*, 1105–1125. [CrossRef] [PubMed]
5. Govindaraj, M.; Huang, W.-C.; Lee, C.-Y.; Lakshmanan, V.; Liu, Y.-H.; So, P.B.; Lin, C.-H.; Chen, J.-D. Structural Diversity of Mercury(II) Halide Complexes Containing Bis-pyridyl-bis-amide with Bulky and Angular Backbones: Ligand Effect and Metal Sensing. *Int. J. Mol. Sci.* **2022**, *23*, 7861. [CrossRef] [PubMed]
6. Lakshmanan, V.; Lai, Y.-T.; Yang, X.-K.; Govindaraj, M.; Lin, C.-H.; Chen, J.-D. Eight-Fold Interpenetrating Diamondoid Coordination Polymers for Sensing Volatile Organic Compounds and Metal Ions. *Polymers* **2021**, *13*, 3018. [CrossRef]
7. Chhetri, P.M.; Wu, M.-H.; Hsieh, C.-T.; Yang, X.-K.; Wu, C.-M.; Yang, E.-C.; Wang, C.-C.; Chen, J.-D. A Pair of CoII Supramolecular Isomers Based on Flexible Bis-Pyridyl-Bis-Amide and Angular Dicarboxylate Ligands. *Molecules* **2020**, *25*, 201. [CrossRef]
8. Thapa, K.B.; Chen, J.-D. Crystal engineering of coordination polymers containing flexible bis-pyridyl-bis-amide ligands. *Crys tEngComm* **2015**, *17*, 4611–4626. [CrossRef]
9. Tang, S.; Gao, Y.; Han, S.; Chi, J.; Zhang, Z.; Liu, G. A kind of complex-based electrocatalytic sensor for monitoring the reduction of Cr(VI) by organic/inorganic reductants. *Polyhedron* **2022**, *223*, 115944. [CrossRef]
10. Liu, G.; Han, S.; Gao, Y.; Xu, N.; Wang, X.; Chen, B. Multifunctional fluorescence responses of phenylamide-bridged d^{10} coordination polymers structurally regulated by dicarboxylates and metal ions. *CrystEngComm* **2020**, *22*, 7952–7961. [CrossRef]
11. Huang, S.-Y.; Li, J.-Y.; Li, J.-Q.; Xu, W.-Y.; Luo, M.-B.; Zhua, Y.; Luo, F. Exceptional temperature-dependent coordination sites from acylamide groups. *Dalton Trans.* **2014**, *43*, 5260–5264. [CrossRef] [PubMed]
12. Blatov, V.A.; Shevchenko, A.P.; Proserpio, D.M. Applied Topological Analysis of Crystal Structures with the Program Package ToposPro. *Cryst. Growth Des.* **2014**, *14*, 3576–3586. [CrossRef]
13. Bonneau, C.; O'Keeffe, M.; Proserpio, D.M.; Blatov, V.A.; Batten, S.R.; Bourne, S.A.; Lah, M.S.; Eon, J.G.; Hyde, S.T.; Wiggin, S.B.; et al. Deconstruction of Crystalline Networks into Underlying Nets: Relevance for Terminology Guidelines and Crystallographic Databases. *Cryst. Growth Des.* **2018**, *18*, 3411–3418. [CrossRef]
14. Wen, T.; Zhang, D.-X.; Zhang, J. Two-Dimensional Copper(I) Coordination Polymer Materials as Photocatalysts for the Degradation of Organic Dyes. *Inorg. Chem.* **2013**, *52*, 12–14. [CrossRef]
15. Wang, X.-L.; Luan, J.; Sui, F.-F.; Lin, H.-Y.; Liu, G.-C.; Xu, C. Structural Diversities and Fluorescent and Photocatalytic Properties of a Series of Cu^{II} Coordination Polymers Constructed from Flexible Bis-pyridyl-bis-amide Ligands with Different Spacer Lengths and Different Aromatic Carboxylates. *Cryst. Growth Des.* **2013**, *13*, 3561–3576. [CrossRef]
16. Wu, Y.-P.; Wu, X.-Q.; Wang, J.-F.; Zhao, J.; Dong, W.-W.; Li, D.-S.; Zhang, Q.-C. Assembly of Two Novel Cd_3/(Cd_3 + Cd_5)-Cluster-Based Metal−Organic Frameworks: Structures, Luminescence, and Photocatalytic Degradation of Organic Dyes. *Cryst. Growth Des.* **2016**, *16*, 2309–2316. [CrossRef]
17. Cheng, H.-J.; Tang, X.-Y.; Yuan, R.-X.; Lang, J.-P. Structural diversity of Zn(II) coordination polymers based on bis-imidazolyl ligands and 5-R-1,3-benzenedicarboxylate and their photocatalytic properties. *CrystEngComm* **2016**, *18*, 4851–4862. [CrossRef]
18. Li, D.-X.; Ren, Z.-G.; Young, D.J.; Lang, J.-P. Synthesis of Two Coordination Polymer Photocatalystsand Significant Enhancement of Their CatalyticPhotodegradation Activity by Doping with Co^{2+} Ions. *Eur. J. Inorg. Chem.* **2015**, *11*, 1981–1988. [CrossRef]
19. Wang, X.-L.; Sha, X.-T.; Liu, G.-C.; Chen, N.-L.; Tian, Y. Polycarboxylate-directed various CoIJII) complexes based on a "V"-like bis-pyridyl-bis-amide derivative: Construction, electrochemical and photocatalytic properties. *CrystEngComm* **2015**, *17*, 7290–7299. [CrossRef]
20. Chang, M.-N.; Yang, X.-K.; Chhetri, P.M.; Chen, J.-D. Metal and Ligand Effects on the Construction of Divalent Coordination Polymers Based on bis-pyridyl-bis-amide and Polycarboxylate Ligands. *Polymers* **2017**, *9*, 691. [CrossRef]
21. Thapa, K.B.; Wu, M.-H.; Yang, X.-K.; Chen, T.-R.; Chen, J.-D. Co(II) coordination polymers exhibiting reversible structural transformation and color change: A comparative analysis with Ni(II) analogue. *Polyhedron* **2018**, *152*, 225–232. [CrossRef]
22. Sumby, C.J.; Hanton, L.R. Syntheses and studies of flexible amide ligands: A toolkit for studying metallo-supramolecular assemblies for anion binding. *Tetrahedron* **2009**, *65*, 4681–4691. [CrossRef]
23. Bruker AXS. *APEX2, V2008.6, SADABS V2008/1, SAINT V7.60A, SHELXTL V6.14*; Bruker AXS Inc.: Madison, WI, USA, 2008.
24. Sheldrick, G.M. A short history of SHELX. *Acta Crystallogr.* **2008**, *A64*, 112–122. [CrossRef] [PubMed]

molecules

MDPI

Article

Composite Materials Based on a Zr^{4+} MOF and Aluminosilicates for the Simultaneous Removal of Cationic and Anionic Dyes from Aqueous Media

Petros Georgianos [1], Anastasia D. Pournara [1], Evangelos K. Andreou [2], Gerasimos S. Armatas [2] and Manolis J. Manos [1,3,*]

1 Department of Chemistry, University of Ioannina, GR-45110 Ioannina, Greece
2 Department of Materials Science and Technology, University of Crete, GR-70013 Heraklion, Greece
3 Institute of Materials Science and Computing, University Research Center of Ioannina, GR-45110 Ioannina, Greece
* Correspondence: emanos@uoi.gr

Abstract: Environmental pollution has been a reality for many decades, with its contamination intensifying daily due to rapid urbanization and the ever-increasing world population. Dyes, and especially synthetic ones, constitute a category of pollutants that not only affect the quality of water but also exhibit high toxicity toward living organisms. This study was thoroughly planned to explore the removal of two toxic dyes, namely the methylene blue (MB) and methyl orange (MO) compounds from contaminated aqueous media. For this purpose, we designed and synthesized two new composite materials based on ammonium-functionalized Zr^{4+} MOF (MOR-1 or UiO-66-NH_3^+) and naturally occurring sorbents, such as bentonite and clinoptilolite. The composite materials displayed exceptional sorption capability toward both MB^+ and MO^- ions. A key finding of this study was the high efficiency of the composite materials to simultaneously remove MB^+ and MO^- under continuous flow conditions, also showing regeneration capability and reusability, thus providing an alternative to well-known mixed bed resins.

Keywords: metal–organic frameworks; composite materials; dyes' sorption; methylene blue; methyl orange; zeolite; clay; sorption column; alginate beads

1. Introduction

In recent centuries, the industrialization of human societies has led to massive urbanization in most regions on planet Earth. Consequently, different kinds of industries have developed, which have served people's needs in various aspects of their daily life. However, the proliferation of people's needs for products and the rapid increase of industrial production have resulted in the generation of hazardous waste and toxic gases, alongside the production of the desired products. Aqueous wastes, originating from industry or from human processes, when released into the environment without any kind of treatment, cause serious water pollution.

Over the years, various pollutants, such as pharmaceuticals and personal care products (PCPs), pesticides, dyes, oil, and aromatic and/or organic compounds, have been detected in aquatic environments [1]. Interestingly, dyes and pigments, which are widely used in textile industries, tanneries, paper mills, and plastics production, are the most abundant pollutants in aquatic environments. It is estimated that approximately 100,000 commercially available dyes are produced each year at a rate of 8×10^5 tons per year, with 15% of the dyes produced typically being released as wastewater to water bodies [2–4]. Water quality is greatly affected by color, and even a small percentage of dye can be visible and cause a color sensation in the water [5]. These seemingly 'innocent pollutants' are harmful to fish and other aquatic organisms, are potentially carcinogenic, and can cause acute or chronic diseases to exposed living beings [6,7]. Moreover, dyes exhibit resistance to light, heat, and oxidizing agents, a fact that is derived from their complex aromatic molecular structures. On that account, the protection of the earth's water resources becomes vital and has been facilitated through the development of effective and efficient separation and purification techniques.

So far, the reported techniques for the treatment of dye-contaminated aquatic sources include sorption, flocculation, electrolysis, photocatalysis, and biodegradation [8–11]. Among these methods, sorption is the simplest and most efficient one adopted for these purposes [12]. Thus, in the last decades, special attention has been paid to the research on nanoporous materials, as it has been found that they are good sorbents and play an important role in separation and purification processes. The relatively new class of highly crystalline porous solids, metal–organic frameworks (MOFs), is constantly gaining ground in various research fields, and especially in applications regarding sorption removal and the separation/purification of contaminated wastewater [13–16].

In our previous work, we reported the successful removal of the acidic/anionic dye methyl orange (MO^-) with the $[Zr_6O_4(OH)_4(NH_3^+\text{-BDC})_6]Cl_6\cdot$solvent (MOR-1 or UiO-66-NH_3^+), where $NH_2\text{-BDC}^{2-}$ = 2-amino-terephthalate, in its composite form with alginic acid (HA) [12]. Although MOR-1-HA has been shown to be an excellent sorbent toward MO^-, the results were rather discouraging when the cationic dye methylene blue (MB^+) was used as the target pollutant, confirming our prediction that the sorption mechanism involves an anion-exchange process [12]. To this end, we decided to improve the sorbent functionality to enable the removing of both anionic and cationic species. Given the fact that MOR-1-HA could easily exchange its Cl^- with MO^-, the goal was to design a composite which consisted of MOR-1 and a secondary unit that could capture cationic species. Clays and zeolites were two promising classes of materials since they have been extensively applied in the removal of cationic dyes from wastewater, indicating superior sorption behavior with great cation-exchange properties [17,18]. Herein, we present two composite materials based on the anion-exchanging material MOR-1, combined with natural inorganic cation-exchange materials, specifically bentonite (clay) and clinoptilolite (zeolite), together with alginic acid—a natural organic polymer. The composite materials **(MOR-1/Bentonite)-HA** and **(MOR-1/Clinoptilolite)-HA** exhibited an exceptional capability to remove MO^- and MB^+ dyes from water solutions under various conditions. Additionally, we synthesized two composite sorbents based on the above-mentioned materials, calcium alginate and Fe_3O_4, which in the form of beads could be successfully employed in ion-exchange columns. It is worth mentioning that such columns combine the capability for the simultaneous reduction of the concentrations of mixed MO^-/MB^+ dyes in aqueous solutions and regeneration capacity reusability, thus offering an alternative to the use of mixed bed resins [19].

2. Results

2.1. Synthesis of the Composite Materials

In our group, several ammonium-functionalized Zr^{4+} MOFs have been reported and widely used as sorbents for the removal of various anionic species [12,20–25]. However, these MOFs could not be applied in solutions containing cationic pollutants without the previous deprotonation of the amine group [26], which means that the material could not further uptake anions. The principal idea in this study was to design composite multifunctional sorbents based on MOR-1 (Figure 1), an excellent anion sorbent, and widely available aluminosilicate materials, such as bentonite and clinoptilolite (Figure S1) that are well-known for their highly efficient cation-exchange properties [27–30]. Thus, the resulted materials were capable of simultaneously removing both anionic and cationic pollutants.

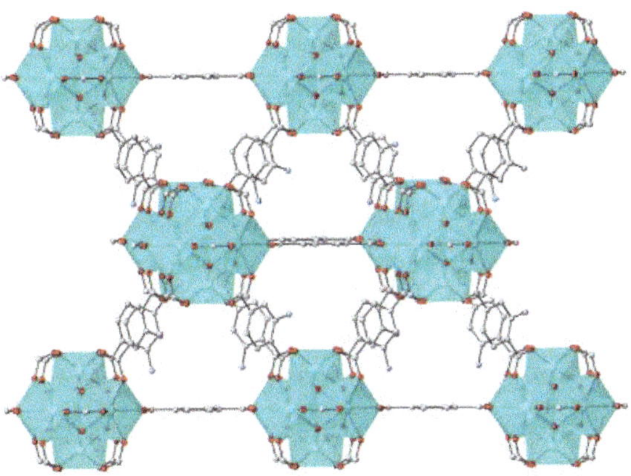

Figure 1. Representation of the structure of MOR-1. Color code: C, grey; O, red; N, blue; Zr, cyan.

The synthesis of these composite materials was carried out in aqueous solutions of sodium alginate where MOR-1 and bentonite or clinoptilolite were mixed in a weight percentage ratio of 1:1. Within these solutions, layers of alginate anions covered the particles of the materials and the addition of concentrated acetic acid led to the precipitation of the **(MOR-1/Bentonite)-HA** and **(MOR-1/Clinoptilolite)-HA (HA = alginic acid)** composites. The weight percentage of alginate's contribution in each composite was only 2% and did not affect the sorption process. These materials were proven excellent sorbents via batch sorption studies (see below). However, they were isolated in the form of fine powder, and thus they were not suitable to be used as stationary phases in columns. To this end, we prepared composite materials, namely **(MOR-1/Bentonite/Fe_3O_4)-10%CA** and **(MOR-1/Clinoptilolite/Fe_3O_4)-10%CA (CA = calcium alginate)** in the form of beads (Figure 2), which contained more alginate content (10%) as well as Fe_3O_4 particles. The latter component not only allowed an easy recovery of the beads by applying an external magnetic field (Figure S2), but also seemed to enhance the mechanical strength of the beads, facilitating the unhindered flow of water solutions through the stationary phase of the column. We should note that the significantly increased alginate content in the beads, compared to that of the powder form of the composites, may result in slower sorption kinetics; however, the isolation of composites in the form of beads requires a content of alginate $\geq$10% *w/w*.

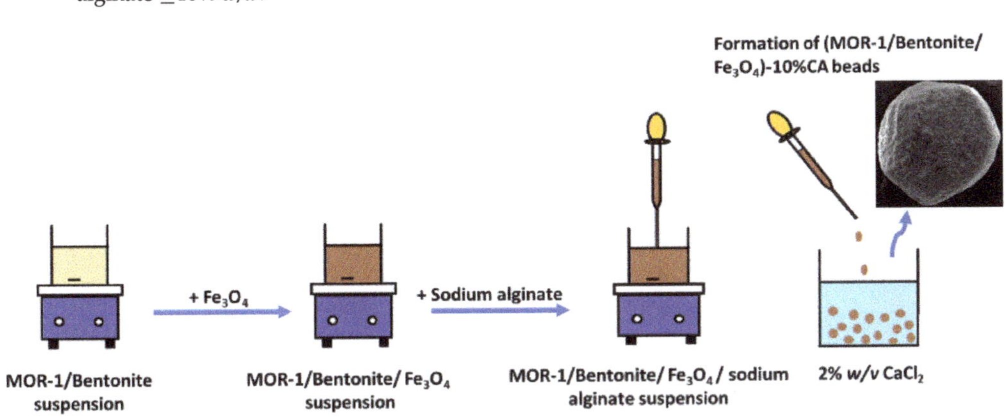

Figure 2. Schematic representation of the preparation of **(MOR-1/Bentonite/Fe_3O_4)-10%CA beads.**

2.2. Characterization of the Composite Materials

In our previous work, we reported that MOR-1 and its alginic acid composite were highly crystalline materials [20,21]. The PXRD pattern of bentonite revealed a typical amorphous structure from 4 to 40 2θ (Figure 3A). In PXRD patterns obtained after the synthesis of **(MOR-1/Bentonite)-HA**, a slight loss of the crystallinity of MOR-1-HA was observed; however, the characteristic diffraction peaks of MOR-1-HA at 2θ of 7.3°, 8.5°, and 12° corresponding to (111), (200), and (220) planes, respectively, were present (Figure 3A). A typical PXRD pattern of clinoptilolite displays several diffraction peaks (Figure 3B) [31]. The PXRD pattern of **(MOR-1/Clinoptilolite)-HA** (Figure 3B) contained diffraction peaks of both MOR-1-HA and clinoptilolite, thus confirming the successful preparation of the composite material. Similarly, the PXRD patterns of Fe_3O_4-modified composites revealed great consistency with those of pristine materials (Figure S3).

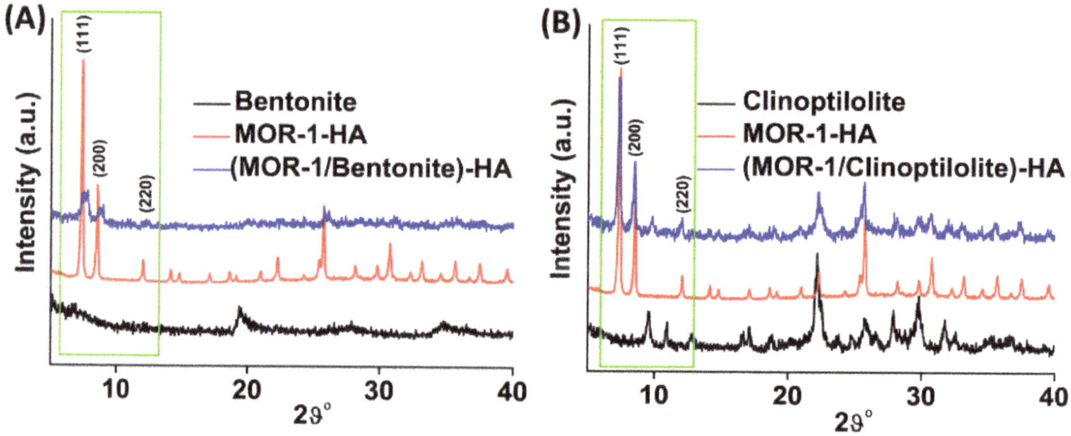

Figure 3. (**A**) PXRD patterns of bentonite, MOR-1-HA, and **(MOR-1/Bentonite)-HA.** (**B**) PXRD patterns of clinoptilolite, MOR-1-HA, and **(MOR-1/Clinoptilolite)-HA. The characteristic diffraction peaks of MOR-1 and aluminosilicates are highlighted in the green rectangles.**

The successful modification of MOR-1 with either bentonite or clinoptilolite was further supported by FT-IR spectroscopy (Figure 4). The IR spectrum of **(MOR-1/Bentonite)-HA** indicated several characteristic bands of MOR-1, such as those at 1565 and 1383 cm^{-1} assigned to the $\nu_{as}(COO^-)$ and $\nu_s(COO^-)$ stretching modes, respectively, as well as the characteristic bands of pristine bentonite. Specifically, the peak at 3627 cm^{-1} was attributed to the vibration of the Al-O-H group, while the peaks at 1117 and 1030 cm^{-1} were due to the vibrations of Si-O units. Moreover, the bands at 524 and 463 cm^{-1} were related to the Al-O-Si and Si-O-Si bending vibrations of bentonite, respectively (Figure 4A) [32]. Similarly, the IR spectrum of **(MOR-1/Clinoptilolite)-HA** revealed the characteristic bands for MOR-1 as well as those for clinoptilolite (Figure 4B). In addition, the IR spectra of **(MOR-1/Bentonite/Fe_3O_4)-10%CA** and **(MOR-1/Clinoptilolite/Fe_3O_4)-10%CA** displayed the characteristic peaks for MOR-1 and either of those for bentonite (Figure S4a) or clinoptilolite materials (Figure S4b).

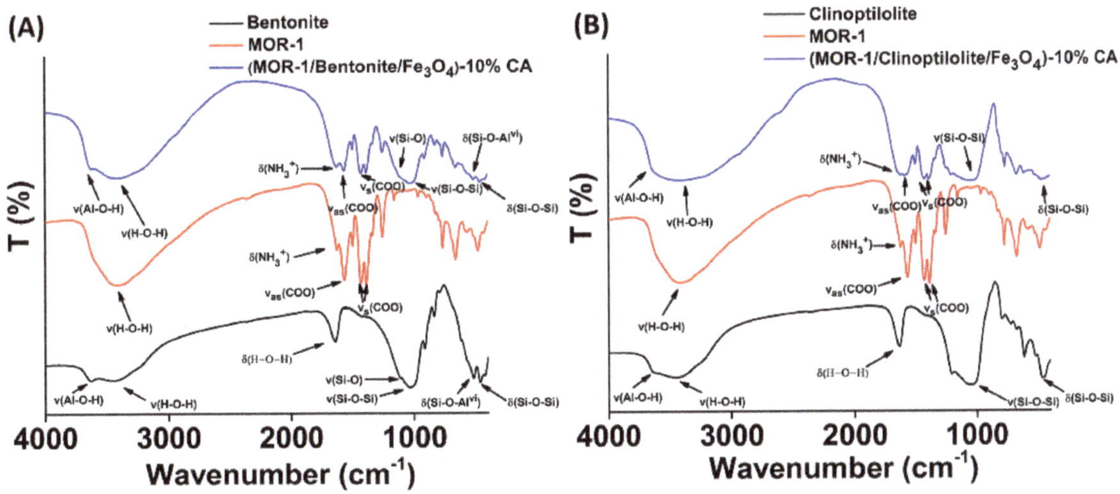

Figure 4. FT-IR spectra of (**A**) bentonite, MOR-1, and **(MOR-1/Bentonite)-HA**; (**B**) clinoptilolite, MOR-1, and **(MOR-1/Clinoptilolite)-HA**.

Furthermore, diffuse reflectance UV-Vis data for the composite materials indicated absorption features of both MOR-1 and aluminosilicate components (Figures S5 and S6). The Brunauer–Emmett–Teller (BET) surface areas for **(MOR-1/Bentonite)-HA** and **(MOR-1/Clinoptilolite)-HA** were found to be 313 and 368 m^2/g, respectively (Figure 5), whereas the BET surface areas of **(MOR-1/Bentonite/Fe_3O_4)-10%CA** and **(MOR-1/Clinoptilolite/Fe_3O_4)-10%CA** were determined to be 246 and 286 m^2/g, respectively (Figure S7). In general, the surface areas of the composite materials were found to be significantly smaller than those of pristine MOR-1 (Figure 5), which implied a partial blockage of the pores of the MOF framework by the aluminosilicate material. Scanning electron microscopy (SEM) images revealed a two-phase morphology for all the composite materials (Figure S8). In addition, EDS analysis confirmed further the amalgamation of the MOF with bentonite or clinoptilolite (Figure S9), as well as the presence of Fe in the **(MOR-1/Bentonite/Fe_3O_4)-10%CA** and **(MOR-1/Clinoptilolite/Fe_3O_4)-10%CA** composite materials (Figures S10 and S11).

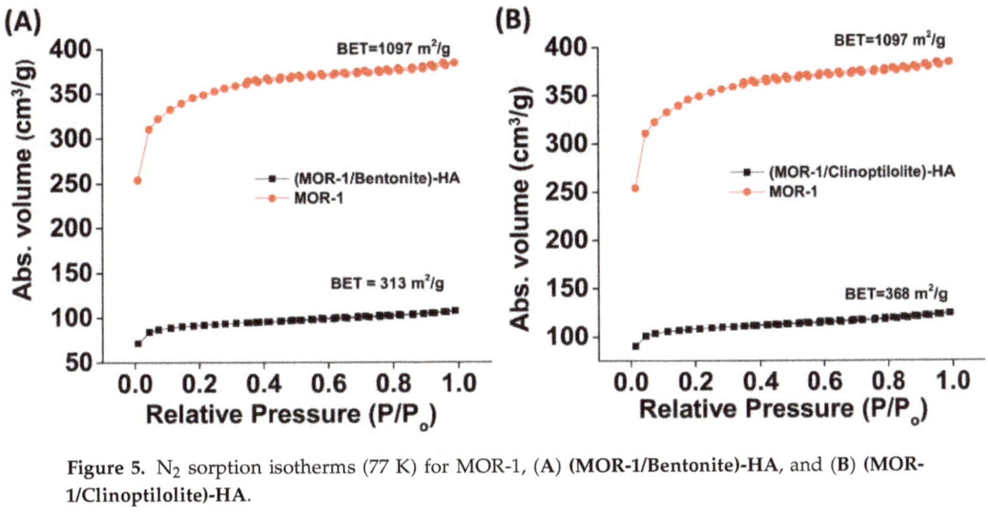

Figure 5. N_2 sorption isotherms (77 K) for MOR-1, (**A**) **(MOR-1/Bentonite)-HA**, and (**B**) **(MOR-1/Clinoptilolite)-HA**.

2.3. Batch Sorption Studies

2.3.1. Sorption Kinetics

The contact time was one of the critical factors that pointedly influenced the performance and assessed the applicability of the sorbent. Thus, the determination of the sorption kinetics was the first step in the investigation of the sorption properties of **(MOR-1/Bentonite)-HA** and **(MOR-1/Clinoptilolite)-HA** toward MB^+ and MO^- ions. Lagergren's first-order equation and Ho–Mckay's pseudo-second-order equation are the commonly used models for the fitting of the kinetics data. In the current study, both models were applied; however, the significantly higher R^2 values, the low values for the residual sum of squares and relatively small uncertainty parameters, suggested that Ho–Mckay's pseudo-second-order equation better represented the kinetics of the dyes' sorption (Table S1). This finding suggested a mechanism involving the chemisorption of dyes on the composite materials [33]. Figure 6 illustrates the effect of the contact time on the sorption of MB^+ and MO^- by the composite materials.

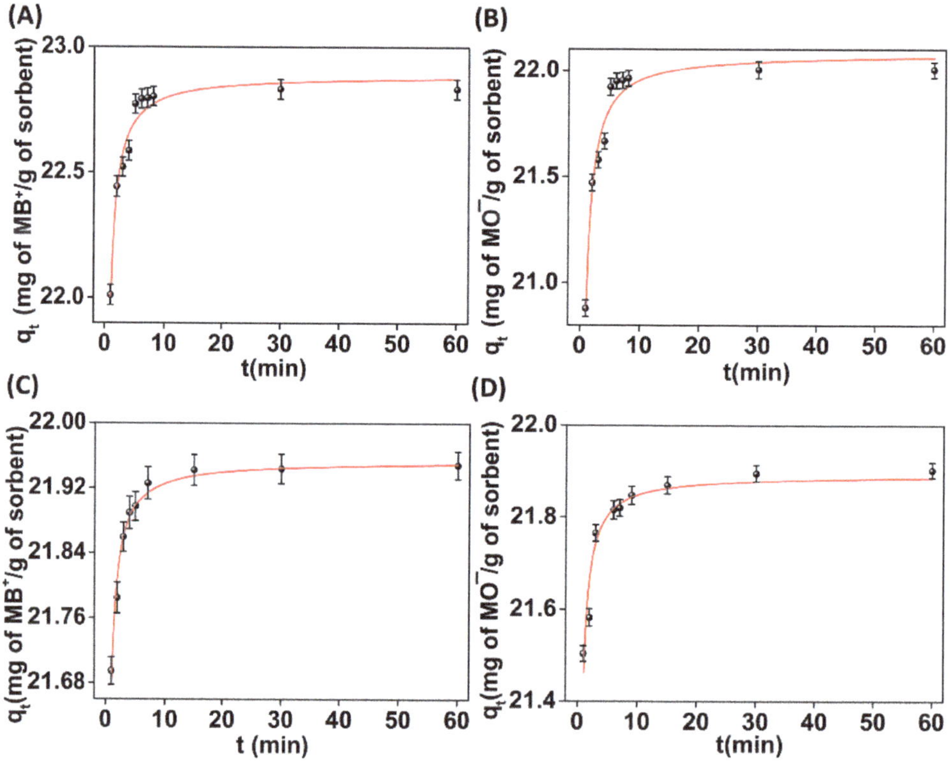

Figure 6. Fitting (red line) of the kinetics data with the Ho–Mckay's pseudo-second-order equation for the sorption of (**A**) MB^+ and (**B**) MO^- by **(MOR-1/Bentonite)-HA**; (**C**) MB^+ and (**D**) MO^- by **(MOR-1/Clinoptilolite)-HA**.

The results indicated that the rate of sorption was very fast in the first 10 min of the interaction between the dyes' ions and the sorbents. After that time, the removal rate became almost insignificant considering that by elapsing the contact time, the vacant sorption sites were less than in the beginning of the experiments. Significantly, within only the 1st min of contact, both materials successfully removed ~97% of the MO^- and ~98% of the MB^+ ions. After 10 min of contact, an equilibrium was reached with removal percentages $\geq$ 99%.

2.3.2. Sorption Isotherms

Equilibrium analysis is a powerful tool for the comprehension of the sorption process. Toward this end, we carried out sorption experiments with solutions of variable concentrations. The sorption equilibrium data were fitted with the Langmuir, Freundlich, and Langmuir–Freundlich isotherm models, the mathematical expressions of which are provided below:

(a) Langmuir

$$q = q_m \frac{bC_e}{1 + bC_e}$$

(b) Freundlich

$$q = K_F C_e^{\frac{1}{n}}$$

(c) Langmuir–Freundlich

$$q = q_m \frac{(bC_e)^{\frac{1}{n}}}{1 + (bC_e)^{\frac{1}{n}}}$$

where q (milligrams per gram) represents the amount of the ion removed at the equilibrium concentration C_e (parts per million), q_m is the maximum sorption capacity of the sorbent, b (milligrams per liter) is the Langmuir constant related to the free energy of the sorption, and K_F and $1/n$ are the Freundlich constants [34].

Figure 7 depicts the equilibrium data for the sorption of MB^+ and MO^- ions by **(MOR-1/Bentonite)-HA** and **(MOR-1/Clinoptilolite)-HA,** as well as their fitting with the suitable model. The sorption isotherm data revealed that **(MOR-1/Bentonite)-HA** was able to remove 321 mg g^{-1} of MB^+ and 312 mg g^{-1} of MO^- from the corresponding solutions. The best fitting of isotherm sorption data for **(MOR-1/Bentonite)-HA** was achieved with the Langmuir–Freundlich model for the sorption of MB^+ and the Langmuir model for the sorption of MO^-. The description of the MO^- sorption data by **(MOR-1/Bentonite)-HA** with the Langmuir model indicated a monolayer sorption of the dyes' anionic species inside the pores of the MOR-1 material [35]. The fact that the MB^+ sorption isotherm of **(MOR-1/Bentonite)-HA** followed the Langmuir–Freundlich model, which combines features of both the Langmuir and Freundlich approaches, indicated the contribution of both homogenous and heterogenous sorption processes [35]. In our previous study, we reported that the sorption of MO^- by MOR-1-HA was attributed to the exchange of Cl^- ions with MO^- ions. However, no sorption capacity was observed for the cationic dye MB^+, which has a similar molecular size to MO^-, revealing that MOR-1–HA material can selectively sorb the anionic but not the cationic dye through an ion-exchange process [12]. These findings suggested that bentonite was responsible for the removal of MB^+ ions and not the MOR-1-HA. Indeed, the latter was supported by the equilibrium data obtained from the sorption of MB^+ by the pristine bentonite material. The maximum sorption capacity was found to be 525 mg g^{-1} (Figure S12a). The value of 321 mg of MB^+ removed per gram of **(MOR-1/Bentonite)-HA** was close to the theoretical, given that the weight percentage of bentonite in the composite material was 50%.

The isotherm sorption data for **(MOR-1/Clinoptilolite)-HA** were fitted with the Langmuir–Freundlich model for both the MB^+ and MO^- sorption processes. The maximum MB^+ and MO^- sorption capacities for **(MOR-1/Clinoptilolite)-HA** were calculated to be 312 mg g^{-1} and 323 mg g^{-1}, respectively. Pristine clinoptilolite was able to remove 403 mg of MB^+ g^{-1} (Figure S12b), while the composite **(MOR-1/Clinoptilolite)-HA** seemed to be more effective for the removal of MB^+. It is worth mentioning that bentonite and clinoptilolite showed no sorption capacity for MO^-. The results acquired from the isotherm studies in combination with those previously obtained for MOR-1-HA [12] revealed that the sorption process involved the cation exchange of MB^+ in the aluminosilicate component and anion exchange of MO^- in the MOR-1 material.

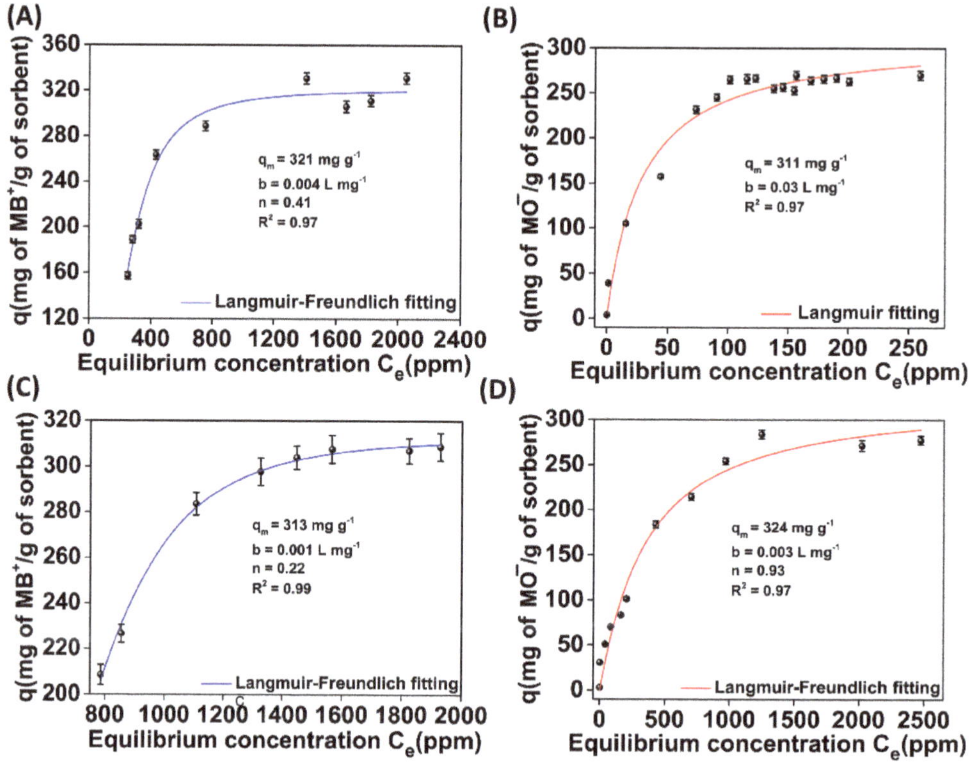

Figure 7. (**A**) MB^+ and (**B**) MO^- isotherm sorption data for **(MOR-1/Bentonite)-HA**; (**C**) MB^+ and (**D**) MO^- isotherm sorption data for **(MOR-1/Clinoptilolite)-HA**.

2.3.3. Variable pH Studies

Detailed studies with dye-contaminated solutions of a wide pH range revealed that pH was not a critical factor for the performance of our sorbents. The pH study was conducted with solutions containing MB^+ or MO^- and in the pH range from 1 to 10 for both materials. Significantly, the capture of MB^+ from either **(MOR-1/Bentonite)-HA** or **(MOR-1/Clinoptilolite)-HA** was overwhelming since the removal percentages were close to 100% (>99.86–100%) independent of the pH of the solution (Figure 8A,C). Furthermore, **(MOR-1/Bentonite)-HA** exhibited astonishing sorptive behavior toward the anionic dye MO^-, with removal percentages higher than 97% (Figure 8B). **(MOR-1/Clinoptilolite)-HA** was found capable to remove MO^- as effectively as **(MOR-1/Bentonite)-HA** in solutions with a pH > 3, while a slight loss was observed at pH 2 and 1, with removal percentages reaching 90% and 72% (Figure 8D), respectively. It is likely that the high excess of Cl^- anions resulted from the HCl acid used for pH adjustment and not that the presence of H^+ affected the sorption of MO^- by (**MOR-1/Clinoptilolite)-HA**. It is not clear, however, why the sorption of MO^- by **(MOR-1/bentonite)-HA** was inhibited only a little under similar conditions.

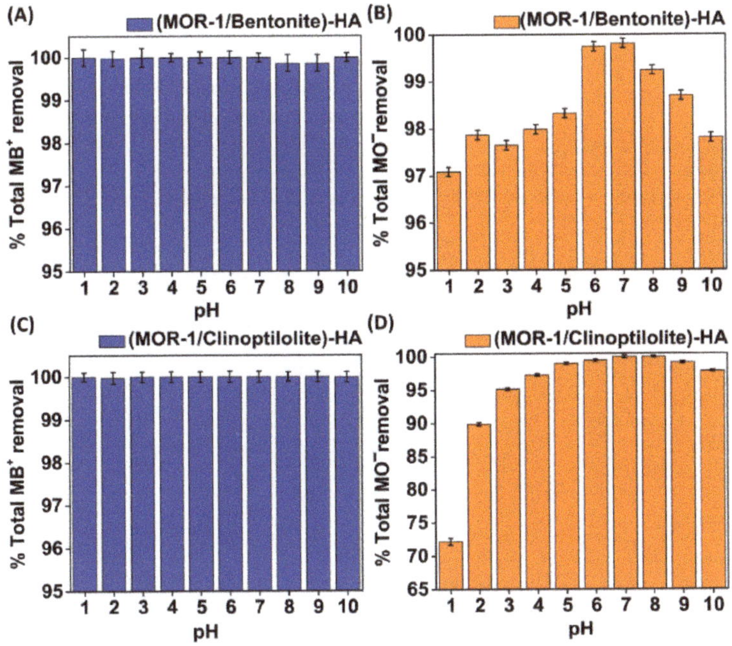

Figure 8. Percentage of sorption of (**A**) MB^+ and (**B**) MO^- by **(MOR-1/Bentonite)-HA**; (**C**) MB^+ and (**D**) MO^- by **(MOR-1/Clinoptilolite)-HA** in the pH range of 1–10.

2.3.4. Selectivity Studies

Given the fact that dye-contaminated wastewater contains several anionic and/or cationic species besides the dyes' ions, dye sorption studies were also performed in complex solutions with a number of antagonistic cations and/or anions. To this end, the sorptive ability of **(MOR-1/Bentonite)-HA** and **(MOR-1/Clinoptilolite)-HA** toward MB^+ was investigated in solutions containing a series of competitive cations, such as Na^+, K^+, and Ca^{2+} in large excesses. Interestingly, the MB^+ removal ability of both composites seemed not to be importantly influenced by the presence of the antagonistic cations. In particular, **(MOR-1/Bentonite)-HA** achieved removal percentages for MB^+ as high as 99.80%, 98.95%, and 99.46% even in the presence of a 1000-fold excess of Na^+, K^+, and Ca^{2+}, respectively, while the correspondent percentages for **(MOR-1/Clinoptilolite)-HA** were found to be 97.90%, 95.40%, and 96.11% (Figure 9A). In addition, the ability of **(MOR-1/Bentonite)-HA** and **(MOR-1/Clinoptilolite)-HA** to capture MO^- was investigated in aqueous solutions containing Cl^-, Br^-, NO_3^-, or SO_4^{2-} anions in relatively high concentrations. As shown in Figure 9B, both composite materials could efficiently remove MO^-, even in the presence of quite high concentrations of competitive anionic species. Specifically, despite the presence of 1000-fold excesses of Cl^- or NO_3^-, the removal percentages of MO^- by **(MOR-1/Bentonite)-HA** and **(MOR-1/Clinoptilolite)-HA** were calculated to be equal or higher than 80%. A slight decrease was observed when a 1000-fold excess of Br^- was added to the solutions (removal percentages = 65.6% and 73.0% for **(MOR-1/Bentonite)-HA** and **(MOR-1/Clinoptilolite)-HA,** respectively), and even higher was the decrease after adding a 1000-fold of SO_4^{2-} (~60% removal). However, the latter finding can be easily explained since SO_4^{2-} is bivalent and can interact more efficiently with the MOF's active sites than the monoanionic MO^-. The last step to the selectivity study was the investigation of the efficiency of **(MOR-1/Bentonite)-HA** and **(MOR-1/Clinoptilolite)-HA** to uptake the ions of dyes under realistic conditions. For this reason, the sorption occurred in bottled water intentionally contaminated with MB^+ or MO^-. These samples were rich in several anionic and cationic species, including Cl^-, NO_3^-, SO_4^{2-}, HCO_3^-, Na^+, K^+, Ca^{2+}, and Mg^{2+}, with concentrations that exceeded those of MB^+ and MO^- by up to 35 times. Although the removal of MO^- seemed to be slightly hindered due to the mixture of the anions, the removal percentages of 67.64% and 65.45% with **(MOR-1/Bentonite)-HA** and **(MOR-1/Clinoptilolite)-HA** were still high and very promising, considering the complexity of these solutions (Figure 9). Moreover, the selective removal of MB^+ by both **(MOR-1/Bentonite)-HA** and **(MOR-1/Clinoptilolite)-HA** was exceptional, since the presence of the ions had zero effect on the sorption process and no traces of the dyes were detected in the bottled samples.

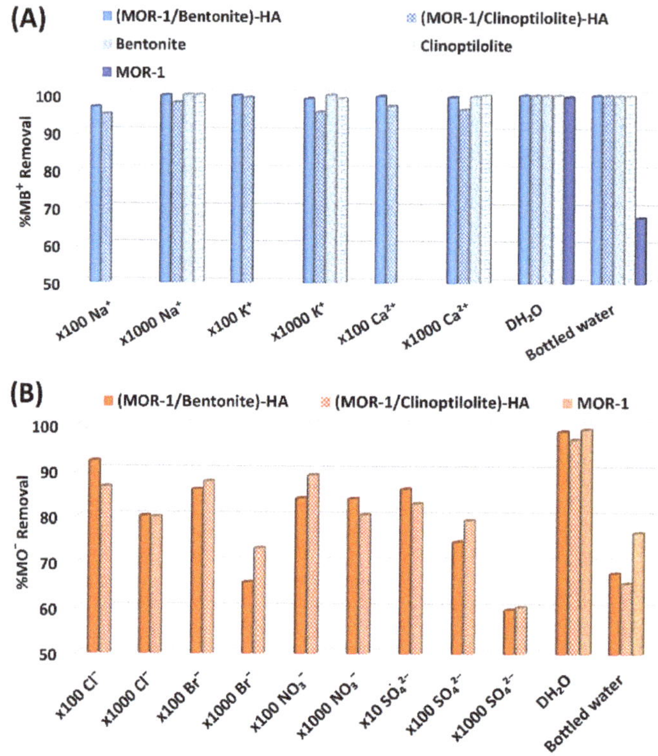

Figure 9. (**A**) MB^+ sorption data for **(MOR-1/Bentonite)-HA** and **(MOR-1/Clinoptilolite)-HA** in the presence of various competitive anions (initial MB^+ concentration = 18 ppm, pH ~ 6.5) and in contaminated bottled water samples (initial MB^+ concentration = 18 ppm, pH ~ 7.8). For comparison, sorption results in distilled water (DH_2O) solutions (containing no antagonistic ions) are also provided. (**B**) MO^- sorption data for **(MOR-1/Bentonite)-HA** and **(MOR-1/Clinoptilolite)-HA** in the presence of various competitive anions (initial MB^+ concentration = 18 ppm, pH ~ 6.5), in contaminated bottled water samples (initial MB^+ concentration = 19 ppm, pH ~ 7.8), and in distilled water solutions. The composition of the bottled water was as follows: pH = 7.8, HCO_3^- = 244 ppm, Cl^- = 4.29 ppm, NO_3^- = 1.93 ppm, SO_4^{2-} = 9.16 ppm, Na^+ = 2.24 ppm, K^+ = 0.6 ppm, Ca^{2+} = 80.7 ppm, and Mg^{2+} = 5.34 ppm.

2.4. Column Sorption Study

The above sorption results derived from the batch reaction experiments are promising for the possible use of **(MOR-1/Bentonite)-HA** and **(MOR-1/Clinoptilolite)-HA** in MO^- and MB^+ sorption applications. However, the efficient performance of ion-exchange materials under stirring conditions does not necessarily mean that they can be applied in industrial wastewater treatment. The latter requires the use of continuous flow ion-exchange columns [19]. Furthermore, dye-contaminated industrial wastewater contains more than one coloring factor, which can be either anionic or cationic. With a view to address this issue, we decided to perform column sorption studies with a mixture of MB^+/MO^-, which better met the requirements for real-world wastewater treatment. As reported above, **(MOR-1/Bentonite)-HA** and **(MOR-1/Clinoptilolite)-HA** are in the form of fine powder, and thus they cannot be used as stationary phases in columns for the removal of dyes under continuous flow. Thus, two new composite materials in the form of beads were isolated and used for the simultaneous removal of a mixture of MB^+/MO^- under dynamic conditions, namely **(MOR-1/Bentonite/Fe_3O_4)-10%CA** and **(MOR-1/Clinoptilolite/Fe_3O_4)-10%CA**. Column sorption studies were conducted with a MB^+/MO^- mixture solution of an initial concentration of 3.7 ppm MB^+ and 5.3 ppm MO^+. Since the light green solution passed through the column, it was decolorized (Figure S13). Importantly, the columns could be easily regenerated by treatment with a solution of 1M HCl and reused for several cycles. Specifically, a column filled with **(MOR-1/Bentonite/Fe_3O_4)-10%CA beads** achieved the removal of 82% of MB^+ from the first bed volume received (bed volume = bed height [cm] x cross sectional area [cm^2]), while after 115 bed volumes, the correspondent removal percentage was close to 48% of the initial MB^+ concentration (Figure S14a). At the same time, 69% and 20% removal were observed for MO^- at the first and last bed volumes collected, respectively (Figure S14b). In the second and third runs, the performance of the column seemed to be improved due to the slower flow rate, which resulted in a longer contact time. The bed volume was 1.15 mL, the average flow rate = 1.66 mL min^{-1}, and the empty bed contact time (EBCT) = 0.69 min [36]. Specifically, in the first bed volume collected in the second run, the removal percentages were increased to 88% and 86% for MB^+ and MO^-, respectively, whereas the corresponding percentages found in the third run of the column were as high as 96% and 90%, respectively (Figure S14). It is worth mentioning that 83% and 75% removal of the initial MB^+ and MO^- content, respectively, could be achieved even after passing 85 bed volumes in the third run of the column. Similar results were obtained from the ion-exchange column filled with **(MOR-1/Clinoptilolite/Fe_3O_4)-10%CA** beads (bed volume = 1.34 mL, the average flow rate = 2.5 mL min^{-1}, and the EBCT = 0.54 min) (Figure S15).

Furthermore, we designed a column sorption investigation where 40 mL of the mixture solution circularly passed through the column. The latter simulated a popular procedure for industrial wastewater treatment, where a series of columns was applied to improve the removal performance of the sorbent [19]. The results proved that both types of columns were impressive, since both **(MOR-1/Bentonite/Fe_3O_4)-10%CA** and **(MOR-1/Clinoptilolite/Fe_3O_4)-10%CA** were able to successfully capture approximately 90% of the initial MB^+/MO^- content (Figure 10). Specifically, **(MOR-1/Bentonite/Fe_3O_4)-10%CA** was able to remove 95% and 92% of the MB^+ and MO^-, respectively, after 10 times of circularly passing the effluent through the column (bed volume = 1.15 mL, the average flow rate = 8 mL min^{-1}, and the EBCT = 0.14 min) (Figure 10A,B). It should be noted that the column was treated with 1M HCl and was reused showing a similar removal capacity. Likewise, the removal percentages for **(MOR-1/Clinoptilolite/Fe_3O_4)-10%CA** were as high as 96% and 90% for MB^+ and MO^-, respectively (bed volume = 1.34 mL, the average flow rate = 8 mL min^{-1}, and the EBCT = 0.17 min) (Figure 10C,D). The regeneration and reuse of this column revealed slight changes in the removal capability of the column. Noteworthily, the average flow rate of the columns was calculated to be 8 mL min^{-1}, which meant that 5 min was enough to complete every cycle of 40 mL solution feeding. The latter finding is very promising for industrial applications, not only because a single column with circularity feeding would be able to sufficiently downgrade a mixture of dye-contaminated wastewater, but also due to the short operation time and the "ready to use" ability after a simple regeneration procedure. In addition, the regeneration of the composite materials was achieved inside the column, in contrast to the traditional mixed bed columns where the regeneration demands removal and repacking at the stationary phase [19].

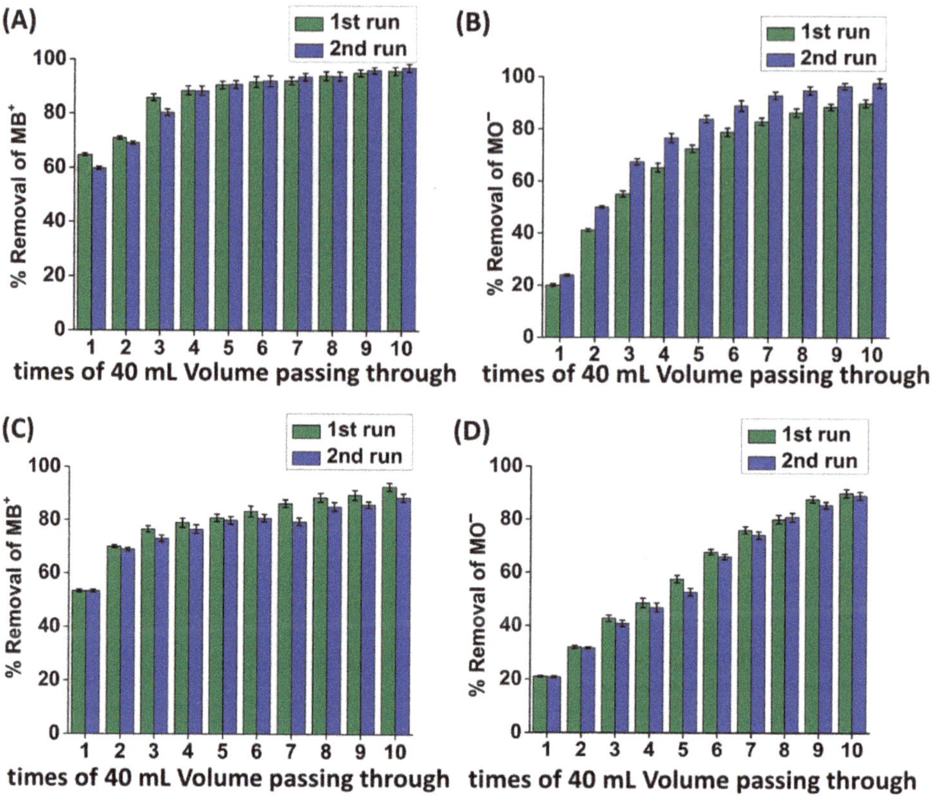

Figure 10. Column sorption data regarding the removal of (**A**) MB^+ and (**B**) MO^- with **(MOR-1/Bentonite/Fe_3O_4)-10%CA** from the mixture solution. Column sorption data regarding the removal of (**C**) MB^+ and (**D**) MO^- with **(MOR-1/Clinoptilolite/Fe_3O_4)-10%CA** from the mixture solution. The MB^+/MO^- mixture solution circularly passed through the column up to 10 times.

Although we cannot clarify if a synergistic effect occurred in the simultaneous sorption of anionic and cationic dyes, several findings from this study supported the opposite scenario. The batch sorption data with solutions containing either MB^+ or MO^- (individual dye sorption experiments) indicated in general the higher sorption capability of composites for MB^+ vs. MO^-. This trend seemed to be followed also in column sorption with a mixture of dyes, and thus synergistic sorption phenomena were not rather likely.

2.5. Isolation and Characterization of the Composite Materials and Dye-Loaded Composite Materials

The color of both composite materials changed from light yellow to blue or orange after the sorption of MB^+ or MO^-, respectively (Figures S16 and S17). Moreover, UV−vis diffuse reflectance spectroscopy further supported the capture of the dyes' ions by the composite materials. Specifically, the wide band appearing in the region of 550 to 800 nm at the spectra of **(MOR-1/Bentonite)-HA@MB^+** (Figure S17a) and **(MOR-1/Clinoptilolite)-HA@MB^+** (Figure S19a) was assigned to the absorption of MB^+. Similarly, in Figures S18b and S19b, the characteristic peak at 464 nm was due to the absorption of MO^-. In addition, UV-vis spectra of the materials used as the stationary phase in the ion-exchange columns displayed absorption peaks of both MB^+ and MO^-, confirming the simultaneous sorption of the two anionic species by the composites (Figure S18). The PXRD patterns of **(MOR-1/Bentonite)-HA@MB^+, (MOR-1/Bentonite)-HA@MO^-, (MOR-1/Clinoptilolite)-**

HA@MB$^+$, and **(MOR-1/Clinoptilolite)-HA@MO$^-$** revealed that the crystal structures were retained after the sorption processes (Figure 11). Moreover, the PXRD patterns obtained from **(MOR-1/Bentonite/Fe_3O_4)-10%CA** and **(MOR-1/Clinoptilolite/Fe_3O_4)-10%CA beads**, after the treatment of the column with the mixture of MB^+ and MO^-, indicated that the crystallinity of the composites was preserved (Figure S21).

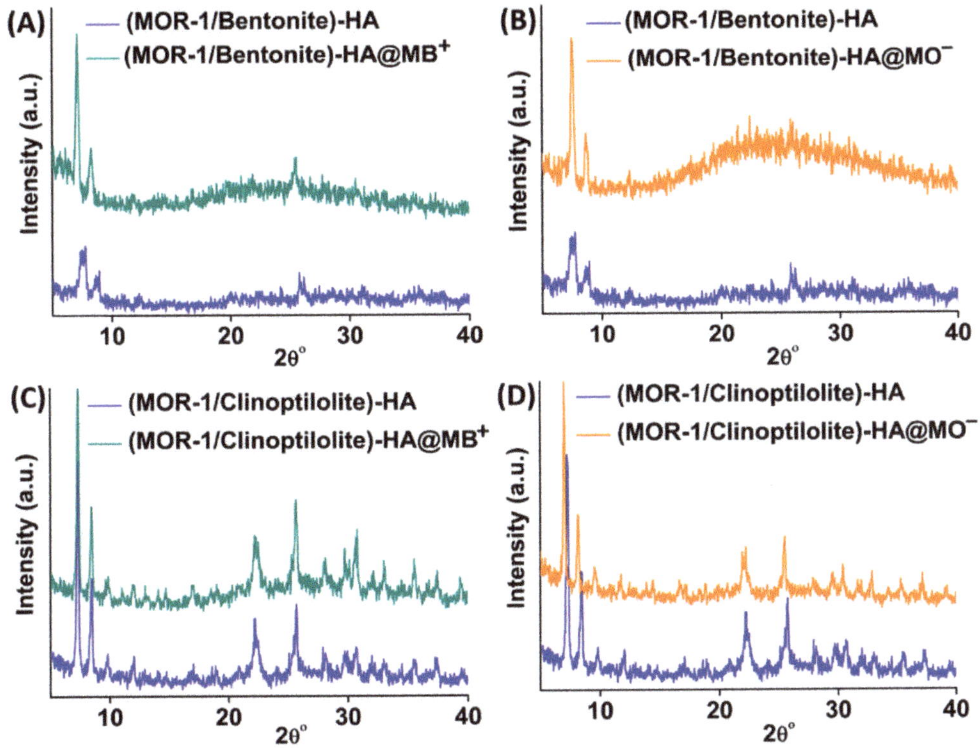

Figure 11. PXRD patterns of **(MOR-1/Bentonite)-HA** along with those of (**A**) **(MOR-1/Bentonite)-HA@MB$^+$**, (**B**) **(MOR-1/Bentonite)-HA@MO$^-$** and **MOR-1/Clinoptilolite)-HA**, (**C**) **(MOR-1/Clinoptilolite)-HA@MB$^+$**, and (**D**) **(MOR-1/Clinoptilolite)-HA@MO$^-$**.

3. Conclusions

In conclusion, this study dealt with the development of novel composite sorbents with the capability for the simultaneous removal of anionic and cationic toxic dyes from aqueous media. Specifically, the two composite materials, **(MOR-1/Bentonite)-HA** and **(MOR-1/Clinoptilolite)-HA**, were synthesized via a facile method and their sorptive efficiencies toward the toxic dyes MB and MO were investigated in detail. **(MOR-1/Bentonite)-HA** and **(MOR-1/Clinoptilolite)-HA** exhibited a high sorption capacity, fast sorption kinetics (the equilibrium can be achieved in ~10 min), excellent sorption ability in acidic and alkaline solutions, and high selectivity for the dyes over various coexisting ionic species. Toward practical applications, **(MOR-1/Bentonite/Fe_3O_4)-10%CA** and **(MOR-1/Clinoptilolite/Fe_3O_4)-10%CA** in the form of beads were used as stationary phases in columns, achieving the highly efficient removal of MB^+ and MO^- ions from a mixture containing both dyes. Overall, this study provides an alternative technology combining the high removal efficiency of multiple pollutants and capability for the in-situ regeneration of sorbents, as opposed to well-known mixed bed columns requiring the ex-situ regeneration of resins, a process that causes significant delays to the water treatment procedure and increased costs.

Supplementary Materials: The supporting information can be downloaded at: https://www.mdpi.com/article/10.3390/molecules28020815/s1, Figure S1: Representation of the structures of bentonite and clinoptilolite; Figure S2: Optical image of (MOR-1/Bentonite/Fe_3O_4)-10%CA beads and recovery of the (MOR-1/Bentonite/Fe_3O_4)-10%CA beads by applying magnetic field; Figure S3: PXRD patterns of bentonite, MOR-1-HA, (MOR-1/Bentonite/Fe_3O_4)-10%CA and clinoptilolite, MOR-1-HA, (MOR-1/Clinoptilolite/Fe_3O_4)-10%CA; Figure S4: FTIR spectra of bentonite, MOR-1, (MOR-1/Bentonite/Fe_3O_4)-10%CA and clinoptilolite, MOR-1, (MOR-1/Clinoptilolite/Fe_3O_4)-10%CA; Figure S5: Solid state UV spectra of bentonite, MOR-1, (MOR-1/Bentonite)-HA and Clinoptilolite, MOR-1, (MOR-1/Clinoptilolite)-HA; Figure S6: Solid state UV spectra of bentonite, MOR-1, (MOR-1/Bentonite/Fe_3O_4)-10%CA and clinoptilolite, MOR-1, (MOR-1/Clinoptilolite/Fe_3O_4)-10%CA; Figure S7: N_2 sorption isotherms (77 K) for MOR-1, (MOR-1/Bentonite/Fe_3O_4)-10%CA and (MOR-1/Clinoptilolite/Fe_3O_4)-10%CA; Figure S8: SEM images of (MOR-1/Bentonite)-HA, (MOR-1/Clinoptilolite)-HA, (MOR-1/Bentonite/Fe_3O_4)-10%CA bead and (MOR-1/Clinoptilolite/Fe_3O_4)-10%CA; Figure S9: EDS spectra of (MOR-1/Bentonite)-HA and (MOR-1/Clinoptilolite)-HA; Figure S10: EDS spectrum of (MOR-1/Bentonite/Fe_3O_4)-10%CA; Figure S11: EDS spectrum of (MOR-1/Clinoptilolite/Fe_3O_4)-10%CA; Figure S12: MB^+ isotherm sorption data for bentonite and clinoptilolite; Figure S13: MB^+/MO^- mixture solution, the colorless effluent of the column and UV spectra of the MB^+/MO^- mixture solution; Figure S14: Column sorption data for (MOR-1/Bentonite/Fe_3O_4)-10%CA beads; Figure S15: Column sorption data for (MOR-1/Clinoptilolite/Fe_3O_4)-10%CA beads; Figure S16: Optical images of (MOR-1/Bentonite)-HA, (MOR-1/Bentonite)-HA@MB^+ and (MOR-1/Bentonite)-HA@MO^-; Figure S17: Optical images of (MOR-1/Clinoptilolite)-HA, (MOR-1/Clinoptilolite)-HA@MB^+ and (MOR-1/ Clinoptilolite)-HA@MO^-; Figure S18: Solid state UV spectra of (MOR-1/Bentonite)-HA in comparison with (MOR-1/Bentonite)-HA@MB^+ and (MOR-1/Bentonite)-HA@MO^-; Figure S19: Solid state UV spectra of (MOR-1/Clinoptilolite)-HA in comparison with (MOR-1/ Clinoptilolite)-HA@MB^+ and (MOR-1/ Clinoptilolite)-HA@MO^-; Figure S20: Solid state UV spectra of (MOR-1/Bentonite/Fe_3O_4)-10%CA and (MOR-1/Clinoptilolite/Fe_3O_4)-10%CA, before and after the sorption; Figure S21: PXRD patterns of (MOR-1/Bentonite/Fe_3O_4)-10%CA and (MOR-1/Clinoptilolite/Fe_3O_4)-10%CA, before and after the sorption; Table S1: The fitting parameters for kinetics. References [37–43] are cited in the Supplementary Materials.

Author Contributions: Investigation, P.G., A.D.P. and E.K.A.; formal analysis, P.G., A.D.P. and E.K.A.; writing—original draft preparation, A.D.P. and M.J.M.; writing—review and editing, A.D.P., G.S.A. and M.J.M.; visualization, M.J.M.; supervision, M.J.M.; project administration, M.J.M.; funding acquisition, M.J.M. All authors have read and agreed to the published version of the manuscript.

Funding: This research project was funded by the Hellenic Foundation for Research and Innovation (H.F.R.I.) under the "1st Call for H.F.R.I. Research Projects to support Faculty Members & Researchers and the Procurement of high-cost research equipment grant" (Project Number: 348).

Institutional Review Board Statement: Not applicable.

Informed Consent Statement: Not applicable.

Data Availability Statement: Not applicable.

Conflicts of Interest: The authors declare no conflict of interest.

Sample Availability: Samples of the compounds are available from the authors.

References

1. Petrie, B.; Barden, R.; Kasprzyk-Hordern, B. A review on emerging contaminants in wastewaters and the environment: Current knowledge, understudied areas and recommendations for future monitoring. *Water Res.* **2015**, *72*, 3–27. [CrossRef]
2. Bhatia, D.; Sharma, N.R.; Singh, J.; Kanwar, R.S. Biological methods for textile dye removal from wastewater: A review. *Crit. Rev. Environ. Sci. Technol.* **2017**, *47*, 1836–1876. [CrossRef]
3. Jamee, R.; Siddique, R. Biodegradation of synthetic dyes of textile effluent by microorganisms: An environmentally and economically sustainable approach. *Eur. J. Microbiol. Immunol.* **2019**, *9*, 114–118. [CrossRef] [PubMed]
4. Yang, C.; Li, L.; Shi, J.; Long, C.; Li, A. Advanced treatment of textile dyeing secondary effluent using magnetic anion exchange resin and its effect on organic fouling in subsequent RO membrane. *J. Hazard. Mater.* **2015**, *284*, 50–57. [CrossRef]

5. Rauf, M.A.; Salman Ashraf, S. Survey of recent trends in biochemically assisted degradation of dyes. *Chem. Eng. J.* **2012**, *209*, 520–530. [CrossRef]
6. Golka, K.; Kopps, S.; Myslak, Z.W. Carcinogenicity of azo colorants: Influence of solubility and bioavailability. *Toxicol. Lett.* **2004**, *151*, 203–210. [CrossRef] [PubMed]
7. Haque, E.; Lo, V.; Minett, A.I.; Harris, A.T.; Church, T.L. Dichotomous adsorption behaviour of dyes on an amino-functionalised metal-organic framework, amino-MIL-101(Al). *J. Mater. Chem. A* **2014**, *2*, 193–203. [CrossRef]
8. Robinson, T.; McMullan, G.; Marchant, R.; Nigam, P. Remediation of dyes in textile effluent: A critical review on current treatment technologies with a proposed alternative. *Bioresour. Technol.* **2001**, *77*, 247–255. [CrossRef] [PubMed]
9. Mezohegyi, G.; van der Zee, F.P.; Font, J.; Fortuny, A.; Fabregat, A. Towards advanced aqueous dye removal processes: A short review on the versatile role of activated carbon. *J. Environ. Manage.* **2012**, *102*, 148–164. [CrossRef]
10. Sarro, M.; Gule, N.P.; Laurenti, E.; Gamberini, R.; Paganini, M.C.; Mallon, P.E.; Calza, P. ZnO-based materials and enzymes hybrid systems as highly efficient catalysts for recalcitrant pollutants abatement. *Chem. Eng. J.* **2018**, *334*, 2530–2538. [CrossRef]
11. Tang, L.; Yu, J.; Pang, Y.; Zeng, G.; Deng, Y.; Wang, J.; Ren, X.; Ye, S.; Peng, B.; Feng, H. Sustainable efficient adsorbent: Alkali-acid modified magnetic biochar derived from sewage sludge for aqueous organic contaminant removal. *Chem. Eng. J.* **2018**, *336*, 160–169. [CrossRef]
12. Pournara, A.D.; Rapti, S.; Skliri, E.; Armatas, G.S.; Tsipis, A.C.; Manos, M.J. Highly Efficient Sorption of Methyl Orange by a Metal–Organic Resin–Alginic Acid Composite. *Chempluschem* **2017**, *82*, 1188–1196. [CrossRef] [PubMed]
13. Li, J.; Wang, X.; Zhao, G.; Chen, C.; Chai, Z.; Alsaedi, A.; Hayat, T.; Wang, X. Metal-organic framework-based materials: Superior adsorbents for the capture of toxic and radioactive metal ions. *Chem. Soc. Rev.* **2018**, *47*, 2322–2356. [CrossRef] [PubMed]
14. Kumar, P.; Pournara, A.; Kim, K.H.; Bansal, V.; Rapti, S.; Manos, M.J. Metal-organic frameworks: Challenges and opportunities for ion-exchange/sorption applications. *Prog. Mater. Sci.* **2017**, *86*, 25–74. [CrossRef]
15. Kang, K.; Liu, S.; Zhang, M.; Li, L.; Liu, C.; Lei, L.; Dai, X.; Xu, C.; Xiao, C. Fast Room-Temperature Synthesis of an Extremely Alkaline-Resistant Cationic Metal–Organic Framework for Sequestering TcO_4^- with Exceptional Selectivity. *Adv. Funct. Mater.* **2022**, *32*, 2208148. [CrossRef]
16. Kang, K.; Shen, N.; Wang, Y.; Li, L.; Zhang, M.; Zhang, X.; Lei, L.; Miao, X.; Wang, S.; Xiao, C. Efficient sequestration of radioactive $^{99}TcO_4^-$ by a rare 3-fold interlocking cationic metal-organic framework: A combined batch experiments, pair distribution function, and crystallographic investigation. *Chem. Eng. J.* **2022**, *427*, 130942. [CrossRef]
17. Kausar, A.; Iqbal, M.; Javed, A.; Aftab, K.; Nazli, Z.I.H.; Bhatti, H.N.; Nouren, S. Dyes adsorption using clay and modified clay: A review. *J. Mol. Liq.* **2018**, *256*, 395–407. [CrossRef]
18. Wang, S.; Peng, Y. Natural zeolites as effective adsorbents in water and wastewater treatment. *Chem. Eng. J.* **2010**, *156*, 11–24. [CrossRef]
19. Zagorodni, A.A. *Ion Exchange Materials: Properties and Applications*; Elsevier: Amsterdam, The Netherlands, 2007; p. 280.
20. Rapti, S.; Pournara, A.; Sarma, D.; Papadas, I.T.; Armatas, G.S.; Tsipis, A.C.; Lazarides, T.; Kanatzidis, M.G.; Manos, M.J. Selective capture of hexavalent chromium from an anion-exchange column of metal organic resin-alginic acid composite. *Chem. Sci.* **2016**, *7*, 2427–2436. [CrossRef]
21. Rapti, S.; Pournara, A.; Sarma, D.; Papadas, I.T.; Armatas, G.S.; Hassan, Y.S.; Alkordi, M.H.; Kanatzidis, M.G.; Manos, M.J. Rapid, green and inexpensive synthesis of high quality UiO-66 amino-functionalized materials with exceptional capability for removal of hexavalent chromium from industrial waste. *Inorg. Chem. Front.* **2016**, *3*, 635–644. [CrossRef]
22. Rapti, S.; Diamantis, S.A.; Dafnomili, A.; Pournara, A.; Skliri, E.; Armatas, G.S.; Tsipis, A.C.; Spanopoulos, I.; Malliakas, C.D.; Kanatzidis, M.G.; et al. Exceptional TcO_4^- sorption capacity and highly efficient ReO_4^- luminescence sensing by Zr^{4+} MOFs. *J. Mater. Chem. A* **2018**, *6*, 20813–20821. [CrossRef]
23. Pournara, A.D.; Rapti, S.; Valmas, A.; Margiolaki, I.; Andreou, E.; Armatas, G.S.; Tsipis, A.C.; Plakatouras, J.C.; Giokas, D.L.; Manos, M.J. Alkylamino-terephthalate ligands stabilize 8-connected Zr^{4+} MOFs with highly efficient sorption for toxic Se species. *J. Mater. Chem. A* **2021**, *9*, 3379–3387. [CrossRef]
24. Pournara, A.D.; Evangelou, D.A.; Roukounaki, C.; Andreou, E.K.; Armatas, G.S.; Lazarides, T.; Manos, M.J. Highly efficient sorption and luminescence sensing of oxoanionic species by 8-connected alkyl-amino functionalized Zr^{4+} MOFs. *Dalton Trans.* **2022**, *51*, 17301–17309. [CrossRef] [PubMed]
25. Pournara, A.D.; Rizogianni, S.; Evangelou, D.A.; Andreou, E.K.; Armatas, G.S.; Manos, M.J. Zr^{4+} -terephthalate MOFs with 6-connected structures, highly efficient As (III/V) sorption and superhydrophobic properties. *Chem. Commun.* **2022**, *58*, 8862–8865. [CrossRef]
26. Diamantis, S.A.; Pournara, A.D.; Koutsouroubi, E.D.; Moularas, C.; Deligiannakis, Y.; Armatas, G.S.; Hatzidimitriou, A.G.; Manos, M.J.; Lazarides, T. Detection and Sorption of Heavy Metal Ions in Aqueous Media by a Fluorescent Zr(IV) Metal–Organic Framework Functionalized with 2-Picolylamine Receptor Groups. *Inorg. Chem.* **2022**, *61*, 7847–7858. [CrossRef]
27. Kubilay, Ş.; Gürkan, R.; Savran, A.; Şahan, T. Removal of Cu(II), Zn(II) and Co(II) ions from aqueous solutions by adsorption onto natural bentonite. *Adsorption* **2007**, *13*, 41–51. [CrossRef]
28. Tahir, S.S.; Rauf, N. Removal of a cationic dye from aqueous solutions by adsorption onto bentonite clay. *Chemosphere* **2006**, *63*, 1842–1848. [CrossRef]
29. Qiu, M.; Qian, C.; Xu, J.; Wu, J.; Wang, G. Studies on the adsorption of dyes into clinoptilolite. *Desalination* **2009**, *243*, 286–292. [CrossRef]

30. Smičiklas, I.; Dimović, S.; Plećaš, I. Removal of Cs^{1+}, Sr^{2+} and Co^{2+} from aqueous solutions by adsorption on natural clinoptilolit *Appl. Clay Sci.* **2007**, *35*, 139–144. [CrossRef]
31. Cadar, O.; Senila, M.; Hoaghia, M.A.; Scurtu, D.; Miu, I.; Levei, E.A. Effects of thermal treatment on natural clinoptilolite-ric zeolite behavior in simulated biological fluids. *Molecules* **2020**, *25*, 2570. [CrossRef]
32. Eren, E.; Afsin, B. An investigation of Cu(II) adsorption by raw and acid-activated bentonite: A combined potentiometri thermodynamic, XRD, IR, DTA study. *J. Hazard. Mater.* **2008**, *151*, 682–691. [CrossRef] [PubMed]
33. Sahoo, T.R.; Prelot, B. *Nanomaterials for the Detection and Removal of Wastewater Pollutants*; Elsevier: Amsterdam, The Netherland 2020.
34. Manos, M.J.; Kanatzidis, M.G. Sequestration of heavy metals from water with layered metal sulfides. *Chem.—A Eur. J.* **2009**, 1 4779–4784. [CrossRef] [PubMed]
35. Wang, J.; Guo, X. Adsorption isotherm models: Classification, physical meaning, application and solving method. *Chemospher* **2020**, *258*, 127279. [CrossRef] [PubMed]
36. Hadi, M.; Samarghandi, M.R.; McKay, G. Simplified fixed bed design models for the adsorption of acid dyes on novel pine con derived activated carbon. *Water Air Soil Pollut.* **2011**, *218*, 197–212. [CrossRef]
37. Karickhoff, S.W.; Bailey, G.W. Optical absorption spectra of clay minerals. *Clays Clay Miner.* **1973**, *21*, 51–57. [CrossRef]
38. Garbowski, E.D.; Mirodatos, C. Investigation of structural charge transfer in zeolites by ultraviolet spectroscopy. *J. Phys. Chen* **1982**, *86*, 97–102. [CrossRef]
39. Garcia-Basabe, Y.; Rodriguez-Iznaga, I.; De Menorval, L.C.; Llewellyn, P.; Maurin, G.; Lewis, D.W.; Binions, R.; Auti M.; Ruiz-Salvador, A.R. Step-wise dealumination of natural clinoptilolite: Structural and physicochemical characterizatio *Microporous Mesoporous Mater.* **2010**, *135*, 187–196. [CrossRef]
40. Zanjanchi, M.A.; Razavi, A. Identification and estimation of extra-framework aluminium in acidic mazzite by diffuse reflectanc spectroscopy. *Spectrochim. Acta Part A Mol. Biomol. Spectrosc.* **2001**, *57*, 119–127. [CrossRef]
41. Bordiga, S.; Buzzoni, R.; Geobaldo, F.; Lamberti, C.; Giamello, E.; Zecchina, A.; Leofanti, G.; Petrini, G.; Tozzola, G.; Vlaic, G. Stru ture and reactivity of framework and extraframework iron in Fe-silicalite as investigated by spectroscopic and physicochemica methods. *J. Catal.* **1996**, *158*, 486–501. [CrossRef]
42. Pérez-Ramírez, J.; Kumar, M.S.; Brückner, A. Reduction of N2O with CO over FeMFI zeolites: Influence of the preparation metho on the iron species and catalytic behavior. *J. Catal.* **2004**, *223*, 13–27. [CrossRef]
43. Benhammou, A.; Yaacoubi, A.; Nibou, L.; Tanouti, B. Adsorption of metal ions onto Moroccan stevensite: Kinetic and isother studies. *J. Colloid Interface Sci.* **2005**, *282*, 320–326. [CrossRef] [PubMed]

molecules

MDPI

Article

In Situ Electrospun Porous MIL-88A/PAN Nanofibrous Membranes for Efficient Removal of Organic Dyes

Hao Wu, Le Xu, Jiao Jia, Fengchun Dong *, Yongtang Jia and Xi Liu *

Guangdong-Hong Kong Joint Laboratory for New Textile Materials, Guangdong Functional Fiber and Textile Engineering Technology Research Center, School of Textile Materials and Engineering, Wuyi University, Jiangmen 529020, China

* Correspondence: dfchun@163.com (F.D.); liuxi_wyu@163.com (X.L.)

Abstract: In recent years, metal–organic framework (MOF)-based nanofibrous membranes (NFMs) have received extensive attention in the application of water treatment. Hence, it is of great significance to realize a simple and efficient preparation strategy of MOF-based porous NFMs. Herein, we developed a direct in situ formation of MOF/polymer NFMs using an electrospinning method. The porous MOF/polymer NFMs were constructed by interconnecting mesopores in electrospun composite nanofibers using poly(vinylpolypyrrolidone) (PVP) as the sacrificial pore-forming agent. MOF (MIL-88A) particles were formed inside the polyacrylonitrile (PAN)/PVP nanofibers in situ during electrospinning, and the porous MIL-88A/PAN (pMIL-88A/PAN) NFM was obtained after removing PVP by ethanol and water washing. The MOF particles were uniformly distributed throughout the pMIL-88A/PAN NFM, showing a good porous micro-nano morphological structure of the NFM with a surface area of 143.21 $m^2\ g^{-1}$, which is conducive to its efficient application in dye adsorption and removal. Specifically, the dye removal efficiencies of the pMIL-88A/PAN NFM for amaranth red, rhodamine B, and acid blue were as high as 99.2, 94.4, and 99.8%, respectively. In addition, the NFM still showed over 80% dye removal efficiencies after five adsorption cycles. The pMIL-88A/PAN NFM also presented high adsorption capacities, fast adsorption kinetics, and high cycling stabilities during the processes of dye adsorption and removal. Overall, this work demonstrates that the in situ electrospun porous MOF/polymer NFMs present promising application potential in water treatment for organic dyestuff removal.

Keywords: electrospun; nanofibrous membranes; MIL-88A; dye adsorption and removal

1. Introduction

Dyes, pesticides, food additives, pharmaceuticals, personal care products, and other emerging organic contaminants in water resources present numerous adverse effects on human health and ecosystem [1]. Owing to the development of industries, such as the textile dyeing and finishing industry, the amount of organic dyes polluting industrial wastewater has increased [1–4]. Therefore, researchers have focused on developing adequate wastewater treatment methods [5,6]. To date, numerous physical and chemical methods, such as advanced oxidation [7], adsorption [8–13], coagulation-flocculation [14], and photocatalysis [15], have been used for organic dye remediation. Of these, adsorption has been one of the most cost-effective methods owing to its simplicity and low energy consumption [16–19]. Therefore, it is critical to develop novel materials with high adsorption capacities for the efficient removal of organic dyes from wastewater.

Metal–organic frameworks (MOFs), owing to their diverse structure, high surface area, and tunable pore size, have been widely used in gas storage and separation applications and organic molecules adsorbents [20–26]. In particular, MOFs with accurate chemical design have been successfully used to adsorb, separate, and remove organic ionic dyes [24,27]. For example, Li et al. prepared a MOF-based nanofiber filter using an electrostatic spinning method and used it to adsorb and selectively separate cationic dyes from aqueous solutions [28]. Jhung et al. prepared urea- or melamine-modified MIL-100(Cr), which is a highly efficient adsorbent for nitroimidazole dyes [29]. However, the applications of powdered MOF particles in organic dye treatment of continuous wastewater are relatively limited [30]. Therefore, it is crucial to develop novel methods for the preparation of polymer–MOF composite membranes for efficient and stable treatment of organic dyes in continuous fluid wastewater.

Electrospinning is a fiber fabrication method used to produce long and continuous polymer fibers with diameters in the nano- or micrometer ranges [31–38]. Furthermore, the fibers form stacks during electrospinning, yielding nonwoven membranes [33,38]. Owing to their high porosity and easy preparation, electrospun polymer fibers have recently been integrated with MOFs to shape MOFs into fibers [29–32]. For example, Zhao et al. proposed a simple method for the uniform growth of porous MIL-100(Fe) on electrospun polyacrylonitrile (PAN) fiber membranes via electrospinning and hydrothermal reactions [38]. Similarly, Lu et al. deposited UiO-66-NH_2 on the surface of electrospun PVDF–MOF linker nanofibrous membranes (NFMs) and used the composite NFMs for removal of toxic chemicals [39]. Xie et al. prepared a graphene oxide (GO)/MIL-88A(Fe) membrane by embedding MIL-88A into a GO matrix. The membrane exhibited remarkable recovery and self-cleaning performance and excellent degradation performance for organic pollutants under ultraviolet (UV) irradiation [40]. Leus et. al. embedded Pt@MIL-101 into a poly-ε-caprolactone matrix via electrospinning to create a "catalytic carpet" [41]. The current methods used to prepare polymer–MOF composite NFMs can be classified as follows: (1) hydrothermal synthesis followed by loading MOFs onto the surface of the NFMs after spinning and (2) direct blending of MOFs with polymer solutions for spinning [28]. These processes are complex; moreover, pure MOF particles are not dispersed uniformly throughout the polymer matrix during spinning [28,42]. Therefore, it is critical to develop new methods to fabricate polymer–MOF composite NFMs.

In this study, we developed a direct in situ formation of MOF/polymer NFM using an electrospinning method. Moreover, the porous MOF/polymer NFMs were constructed by interconnecting mesopores in electrospun composite nanofibers using poly(vinylpolypyrrolidone) (PVP) as the sacrificial pore-forming agent. We designed and prepared MIL-88A powder and developed a simple one-step electrospinning method for in situ binding MIL-88A onto PAN nanofibers. Specifically, MIL-88A MOF particles were formed inside the PAN/PVP fibers in situ during electrospinning, and porous MIL-88A/PAN (pMIL-88A/PAN) NFM was obtained after removing PVP by ethanol and water washing. The MOF particles were uniformly distributed throughout the pMIL-88A/PAN NFM, showing a good porous micro-nano morphological structure of the NFM, which is conducive to its efficient application in dye adsorption and removal. Accordingly, the pMIL-88A/PAN NFM presented high adsorption capacities, fast adsorption kinetics, and high cycling stability for the removal of amaranth red, rhodamine B, and acid blue dyes. Overall, this work provides an effective strategy for electrospinning composite MOF/polymer NFMs, which will benefit future research and applications in water treatment and other fields.

2. Results and Discussion

2.1. Preparation and Characterization of the pMIL-88A/PAN NFMs

Figure 1 shows the schematic for the preparation of the porous MIL-88A/PAN (pMIL-88A/PAN) NFM. Considering the simplicity and universality of the MOF structure, an iron-based MOF (MIL-88A) formed from FuA and $FeCl_3$ $6H_2O$ was used in this study. In situ MOF formation, polymer solidification, and solvent evaporation occurred simultaneously during electrospinning of the spinning solution containing PAN, PVP, FuA, and $FeCl_3$ $6H_2O$. Owing to the difference in solubility, PVP was added to the PAN spinning solution as the supporting pore-forming sacrificial agent, whereas PAN served as the polymer nanofiber skeleton. Therefore, the MIL-88A combined porous nanofiber can be formed after the PVP is removed by ethanol and water washing. More importantly, the originally embedded MIL-88A particles can be regularly exposed to the surface of the porous PAN nanofibers, increasing its specific surface area, and providing abundant adsorption sites for dyes. Meanwhile, the macropores of the PAN fibers provide ample space for adsorption and decrease the mass transfer resistance for dye treatment of continuous wastewater. Overall, the proposed in situ electrospun scheme is a simple and effective strategy for the preparation of pMIL-88A/PAN NFM, and the NFM will be effectively used for the removal of organic dyes in the treatment of continuous wastewater.

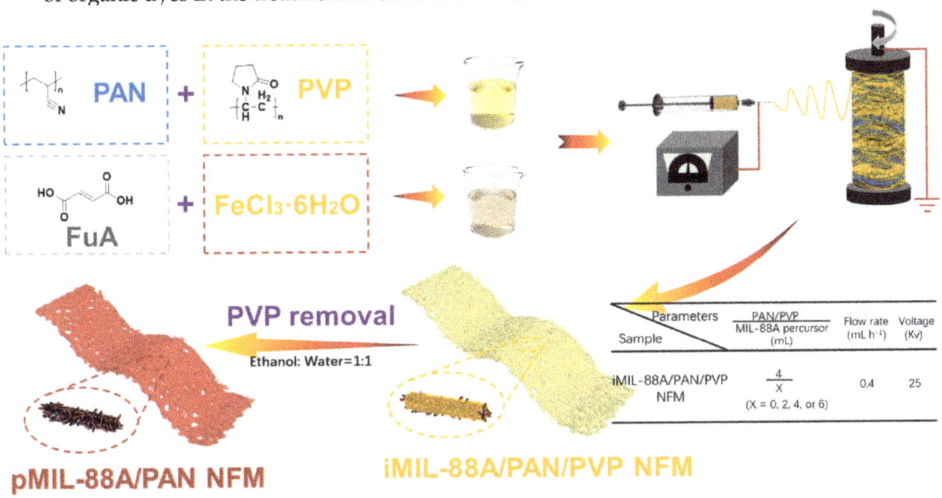

Figure 1. Schematic for the fabrication of pMIL-88A/PAN NFMs.

Details of the preparation of the relevant electrospun NFMs are shown in the Experimental section and Supporting Information. To confirm the formation of the NFMs, their morphology was evaluated. Figure S1a,b show the scanning electron microscopy (SEM) images of the initial PAN composite PVP (PAN/PVP) NFM, presenting uniform nanofibers with diameters of ≈300 nm as well as relatively smooth surfaces, which indicates that PAN and PVP are well mixed and fibrillar during spinning. After sufficient ethanol and water washing of PAN/PVP NFM to remove PVP, porous PAN (pPAN) NFM was obtained. Figure S1c,d shows its rough and uneven surface morphology after removing PVP, suggesting the feasibility of PVP as a supporting pore-forming sacrifice agent for the preparation of porous NFM. Then, FuA and $FeCl_3$ $6H_2O$ were added to PAN/PVP spinning solution to prepare in situ MIL-88A/PAN/PVP (iMIL-88A/PAN/PVP) and pMIL-88A/PAN NFMs. Since the MIL-88A particles formed in situ in the process of electrospinning are wrapped inside the fibers, iMIL-88A/PAN/PVP shows morphological characteristics similar to common nanofibers (Figure 2a,b). Therefore, the aggregation morphology of MIL-88A formed in situ can be observed in SEM images of iMIL-88A/PAN/PVP with removal of PVP, and it can be optimized by adjusting the feeding rate during electrospinning. As shown in Figure S2, less obvious MOF particles were displayed on the surface of NFM at a feeding rate as high as 0.8 mL h^{-1}. With the feeding rate as low as 0.2 mL h^{-1}, however, large prismatic crystal particles of MIL-88A were formed, rendering the NFM fragile, which was not conducive to further application. Through comparison, pMIL-88A/PAN NFM with more suitable morphological characteristic was achieved at a feeding rate of 0.4 mL h^{-1} than 0.6 mL h^{-1} (Figure S2b,c). Correspondingly, Figure 2c,d show the SEM images of the optimal pMIL-88A/PAN NFM with scale bars of 400 nm and 1 μm, respectively. The particles on the nanofiber membrane exhibited long hexagonal rhombic columnar crystal characteristics (Figure S3), indicating that the nanofiber membrane contains a large amount of MIL-88A particles [43–46]. Numerous homogeneous and dense particles with diameters of 50–100 nm appeared on the rough surface of fibers, confirming that MIL-88A was successfully embedded onto the surface of the PAN fibers and formed pMIL-88A/PAN nanofibers. For the purpose of comparison, in situ MIL-88A/PAN (iMIL-88A/PAN) and blended MIL-88A with PAN spinning solution (bMIL-88A/PAN) NFMs were also prepared. As shown in Figure 2e,f, the conventional and relatively smooth surface morphologies indicated that the in situ or pure MOF particles were both encapsulated inside of the PAN fibers, suggesting the significance of PVP auxiliary preparation of pMIL-88A/PAN NFM. Energy-dispersive spectroscopy (EDS) experiments were used to determine the C, N, O, and Fe distributions of the pMIL-88A/PAN NFM, and the EDS spectrum is shown in Figure S4. According to the elemental mappings of the NFM, C, N, O and Fe were well distributed throughout the surface of pMIL-88A/PAN nanofibers (Figure 2g–k), which verified the fabrication and uniform distribution of MIL-88A on the surface of PAN nanofibers. Overall, these results demonstrated that a good micro-nano morphological pMIL-88A/PAN NFM was successfully prepared by electrospinning.

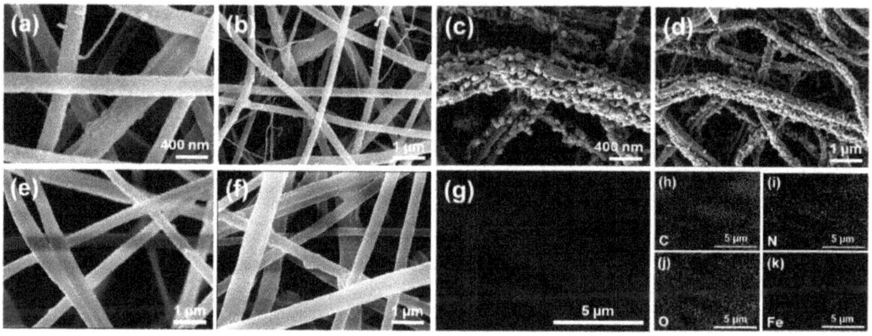

Figure 2. SEM images of the iMIL-88A/PAN/PVP (**a**,**b**) and pMIL-88A/PAN (**c**,**d**) NFMs with scale bars of 400 nm and 1 μm. SEM images of the iMIL-88A/PAN (**e**) and bMIL-88A/PAN (**f**) NFMs. Overall (**g**), C (**h**), N (**i**), O (**j**), and Fe (**k**) elemental mappings of a pMIL-88A/PAN NFM.

The chemical structures of the as-fabricated NFMs were analyzed using FT-IR spectroscopy (Figure 3a,b). As shown in Figure 3a, the characteristic peaks at 2934, 2240, 1730, and 1450 cm^{-1} were assigned to the vibrations of the main chain and pendant group (−C≡N) of PAN, and the carbonyl and tertiary amine groups of PVP were observed with the characteristic peaks of 1660 and 1280 cm^{-1}, respectively. Compared to the PAN/PVP NFM, the absence of characteristic peaks of PVP in the FT-IR spectrum of pPAN NFM indicates that PVP has been effectively and completely removed during the preparation of the porous NFMs. The similar changes of characteristic peaks of PVP in the FT-IR spectra of iMIL-88A/PAN/PVP and pMIL-88A/PAN NFMs are exhibited in Figure 3b, combined with the characteristic peaks of MIL-88A (1608, 1392, and 810–510 cm^{-1}), which confirmed the formation of the chemical structure of pMIL-88A/PAN NFM. The XRD patterns of MIL-88A, FuA, iMIL-88A/PAN/PVP, and pMIL-88A/PAN NFMs are shown in Figure 3c. The peaks at 10.2° and 12.8° in the XRD patterns were ascribed to the (101) and (002) crystal planes of MIL-88A [42], which were present in the XRD patterns of iMIL-88A/PAN/PVP and pMIL-88A/PAN NFMs, indicating that MIL-88A particles were incorporated into the NFMs. The peak at 28.5° of iMIL-88A/PAN/PVP NFM may be attributed to the redundant FuA in the NFM, whereas this peak in pMIL-88A/PAN NFM disappeared due to ethanol and water rinsing of iMIL-88A/PAN/PVP NFM. Furthermore, the intensity of the peak at 10.2° in the XRD pattern of pMIL-88A/PAN NFM was higher than that of iMIL-88A/PAN/PVP NFM, indicating that MIL-88A particles became exposed after washing and a more regular MOF crystal morphology was formed. The above results confirmed the necessity of PVP as a pore forming sacrificial agent, and demonstrated the successful preparation of the pMIL-88A/PAN NFM.

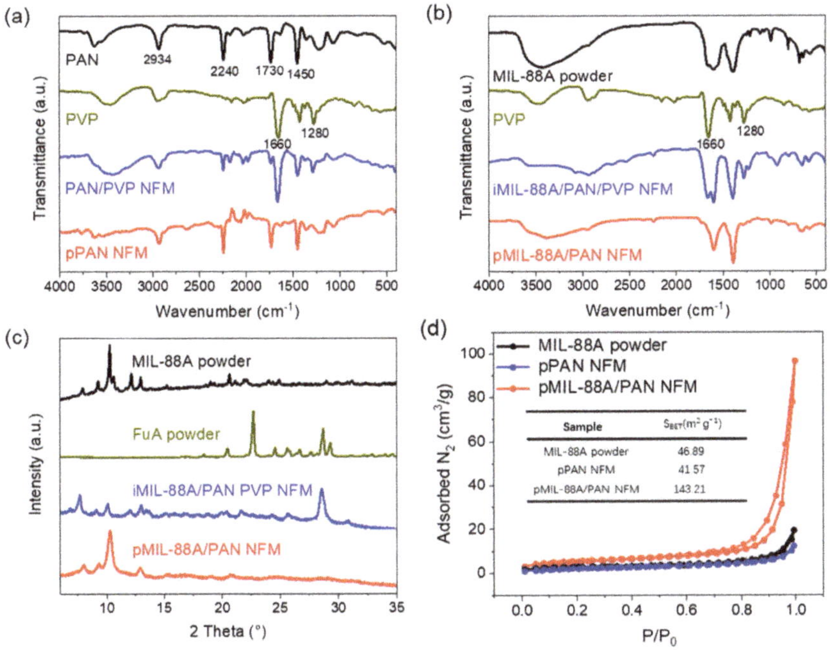

Figure 3. FT−IR spectra (**a**,**b**) and XRD patterns (**c**) of the relevant samples. (**d**) N_2 adsorption−desorption isotherms of the MIL-88A powder, pPAN and pMIL-88A/PAN NFMs; BET surface areas in the inset.

The porosity of the as-fabricated NFMs, which is critical for dye adsorption and removal, was evaluated using N_2 adsorption–desorption experiments. As shown in Figure 3d, the Brunauer–Emmett–Teller (BET) surface area of the pMIL-88A/PAN NFM was calculated to be 143.21 $m^2\ g^{-1}$, which is 3–3.5 times that of MIL-88A powder and pPAN NFM (46.89 and 41.57 $m^2\ g^{-1}$). Moreover, compared to MIL-88A powder and pPAN NFM, the pMIL-88A/PAN NFM showed a high content of pore size distribution in the range of 2–30 nm (Figure S5). These results indicate that the combinations of MIL-88A and porous PAN nanofibers generated abundant mesoporous structures, which were attributed to the formation of MOF–polymer interfaces. Therefore, pMIL-88A/PAN NFM with a good porous structure is expected to achieve efficient dye adsorption and removal.

2.2. Adsorption Performance of the NFMs

2.2.1. Dye Removal Efficiencies

To evaluate the adsorption performance of the as-fabricated pMIL-88A/PAN NFM for organic dyes, amaranth red (AR), rhodamine B (RB), and acid blue (AB) were used as pollutants. These are the most commonly used dyes in industry and are frequently detected in water environments. The chemical structures of the three dyes are shown in Figure 4a. The adsorption performance of the pMIL-88A/PAN NFM was evaluated by removing the dyes from wastewater samples. Figure 4b shows photographs of the dye solutions before and after adsorption and Figure 4c shows the photographs of the NFMs before and after adsorption. The quantified adsorption capacities of the NFMs for AR, RB, and AB were recorded by UV–vis absorption measurements (Figure 4d–f). The ratio of the maximum absorption peak intensity of the solution after and before adsorption can be defined as the dye removal efficiency. The dye removal efficiencies of AR, RB, and AB using pMIL-88A/PAN NFM were calculated as 99.2, 94.4, and 99.8%, respectively. Accordingly, the colors of the three dye solutions became transparent after filtration using the pMIL-88A/PAN NFM, and the colors of the relevant NFMs changed from the initial orange to the corresponding colors of the removed dyes (Figure 4b,c). However, after filtration using the pPAN NFM, the dye removal efficiencies for AR, RB, and AB were calculated as 9.0, 31.8, and 8.6%, respectively, demonstrating that pPAN NFM exhibited limited dye adsorption and removal ability. These results verify that the pMIL-88A/PAN NFM can highly effectively remove organic dyes from aqueous solutions.

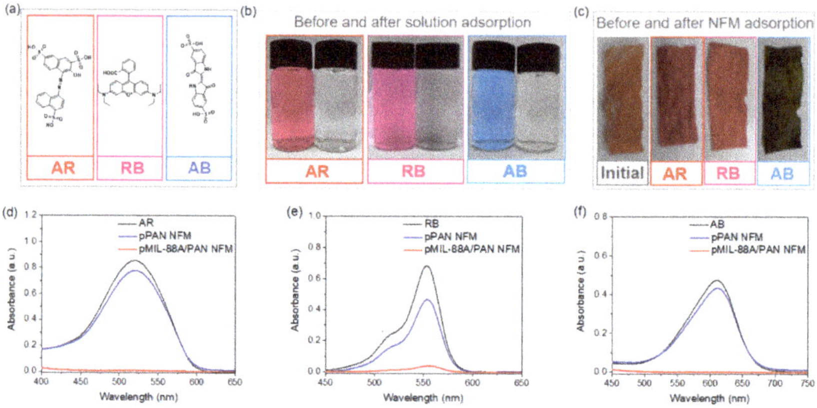

Figure 4. (**a**) Chemical structures of AR, RB, and AB. (**b**) Photographs of dye solutions before and after adsorption. (**c**) Photographs of pMIL-88A/PAN NFMs before and after adsorption. UV−vis absorption spectra of AR (**d**), RB (**e**), and AB (**f**) solutions before and after adsorbed using pPAN and pMIL-88A/PAN NFMs.

2.2.2. Adsorption Kinetics of pMIL-88A/PAN NFM

As shown in Figure 5a,b, due to the abundant adsorption sites of pMIL-88A/PAN NFM, the adsorption rates of the dyes were fast in the initial stages and then started to slow down until equilibriums were reached. The experimental conditions were as follows: C_0 = 20 mg L^{-1} and adsorbent dosage = 0.2 mg mL^{-1}. The kinetic data were subsequently fitted using the most widely used kinetic models: the pseudo-first- and pseudo-second-order kinetic models (Figure 5a and b, respectively). The linear equations of the pseudo-first- and pseudo-second-order kinetic models can be expressed as follows [47]:

$$\log(q_e - q_t) = \log q_e - \frac{k_1 t}{2.303} \quad (1)$$

$$\frac{t}{q_t} = \frac{1}{k_2 q_e 2} + \frac{t}{q_e} \quad (2)$$

where q_t and q_e (mg g^{-1}) are the adsorption capacity at time t and at equilibrium, respectively, and k_1 (h^{-1}) and k_2 (g (h mg)$^{-1}$) are the pseudo-first- and pseudo-second-order model rate constants, respectively. The fitting results are listed in Table 1. The R^2 values revealed that the adsorption processes were better described by the pseudo-second-order kinetic model. Accordingly, the $q_e s$ for the adsorption of AR, RB, and AB using pMIL-88A/PAN NFM were calculated as 102.56, 97.56, and 101.94 mg g^{-1}, respectively, under the pseudo-second-order kinetic model, which is 2.4–4.2 times that of the results calculated with the pseudo-first-order model (23–43 mg g^{-1}).

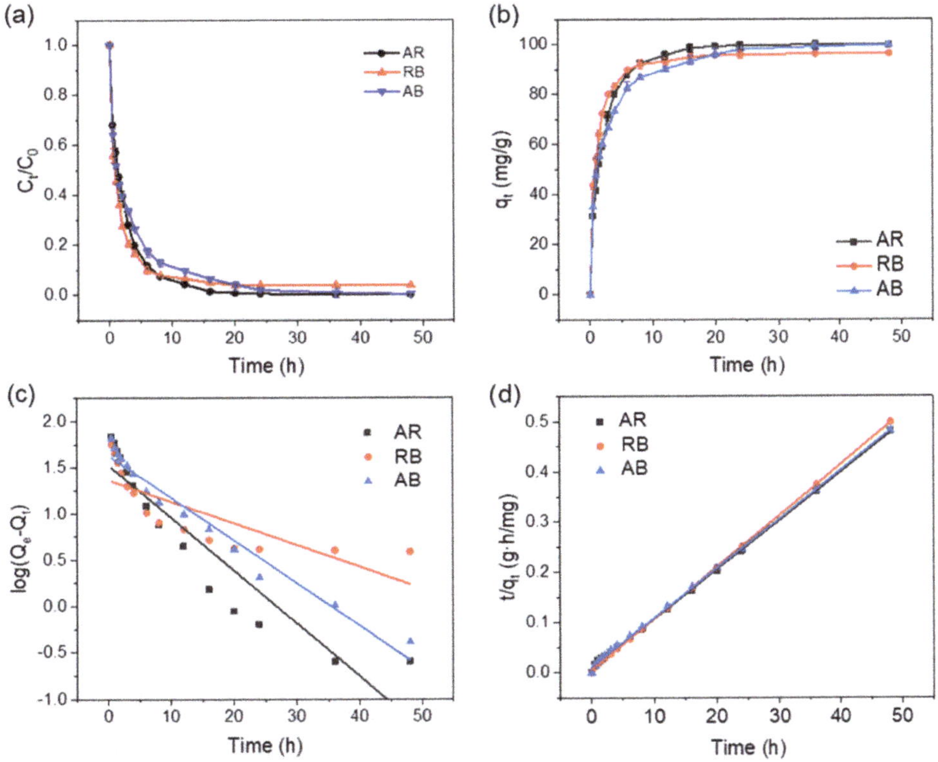

Figure 5. Dye removal rates (**a**) and adsorption kinetics (**b**) of pMIL-88A/PAN NFM. The average fits under pseudo−first (**c**) and pseudo−second−order (**d**) kinetic models for the adsorption of AR, RB, and AB using pMIL-88A/PAN NFM.

Table 1. Calculated results of pseudo-first- and pseudo-second-order constants for the adsorption of AR, RB, and AB using pMIL-88A/PAN NFM.

Dye	Pseudo-First-Order Model			Pseudo-Second-Order Model		
	q_e (mg g^{-1})	k_1 (h^{-1})	R^2	q_e (mg g^{-1})	k_2 (g (mg h)$^{-1}$)	R^2
AR	33.81	0.1329	0.8646	102.56	0.009564	0.9992
RB	23.14	0.0546	0.6184	97.56	0.01846	0.9998
AB	42.88	0.1073	0.9626	101.94	0.008094	0.9992

The dye removal rates for the adsorption of AR, RB, and AB using the pPAN and pMIL-88A/PAN NFMs, and pure MIL-88A powder (Figure S6) indicate that the removal rates of the pMIL-88A/PAN NFM for the three dyes were higher than that of pure MIL-88A powder and pPAN NFM. This was attributed to the synergistic effect of the mesopores and MIL-88A particles morphology of the pMIL-88A/PAN NFM increasing the adsorption efficiency of the composite membranes for the dyes.

2.2.3. Adsorption Isotherms of pMIL-88A/PAN NFM

The adsorption performance of the pMIL-88A/PAN NFM was further evaluated using adsorption isotherms, and the results are shown in Figure 6 and Table 2. The classical isotherm models: Langmuir and Freundlich were used to fit the experimental data, and the corresponding linear equations are as follows [48]:

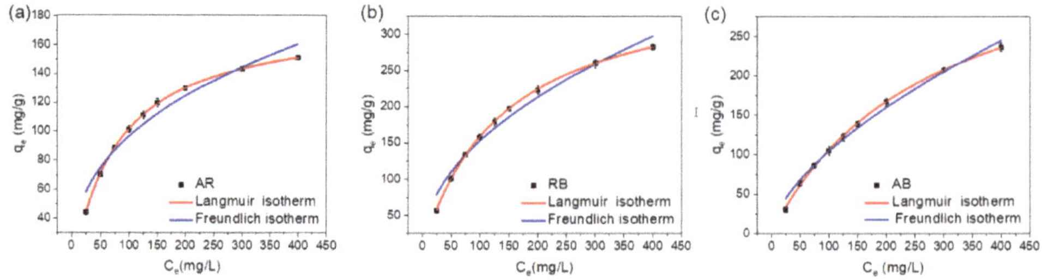

Figure 6. Adsorption isotherms of pMIL-88A/PAN NFM for AR (**a**), RB (**b**) and AB (**c**).

Table 2. Calculated results of the Langmuir and Freundlich models for the adsorption of AR, RB, and AB by the pMIL-88A/PAN NFM.

Dye	Langmuir Isotherm			Freundlich Isotherm		
	q_{max} (mg g^{-1})	b (L mg^{-1})	R^2	K_F	n	R^2
AR	180.41	0.0129	0.99989	17.8564	2.729	0.94561
RB	382.75	0.0071	0.99988	16.6871	2.079	0.97013
AB	399.35	0.0036	0.99950	6.10945	1.711	0.98950

Langmuir isotherm (homogeneous and monolayer adsorption):

$$\frac{C_e}{q_e} = \frac{C_e}{q_m} + \frac{1}{bq_m} \tag{3}$$

Freundlich isotherm (heterogeneous and multilayer adsorption):

$$\log q_e = logK_F + \frac{1}{n}logC_e \tag{4}$$

Here, q_e is the equilibrium adsorption capacity (mg g^{-1}), C_e is the equilibrium concentration (mg L^{-1}), q_m and b are Langmuir constants related to the maximum adsorption capacity and binding energy, respectively, and K_F and n are the Freundlich constant and heterogeneity factor, respectively. The fitting data are summarized in Table 2. Upon analyzing the nonlinear or linear fitting curves (Figure 6) and comparing the R^2 values for the Freundlich and Langmuir models, it was concluded that the AR, RB, and AB adsorption by the pMIL-88A/PAN NFM fit the Langmuir model better. These results indicate that adsorption was a monolayer process, and specific homogeneous sites were present in the adsorbent. Therefore, the maximum adsorption capacities ($q_{max}s$) of the pMIL-88A/PAN NFM for AR, RB, and AB were determined using the Langmuir model, and the results are summarized in Table 2. Meanwhile, adsorption isotherms of MIL-88A powder, iMIL-88A/PAN and bMIL-88A/PAN NFMs for three dyes were performed (Figure S7), and the corresponding $q_{max}s$ are listed in Table S1. The $q_{max}s$ of pure MIL-88A for AR, RB, and AB were 114.03, 119.30, and 309.45 mg g^{-1}, respectively. In contrast, the $q_{max}s$ of the pMIL-88A/PAN NFM for AR, RB, and AB were 180.41, 382.75, and 399.35 mg g^{-1}, respectively. The $q_{max}s$ of the pMIL-88A/PAN NFM for the three dyes, especially for RB, were higher than those of pure MIL-88A powder. In addition, the $q_{max}s$ of the iMIL-88A/PAN and bMIL-88A/PAN NFMs for the three dyes were substantially lower than those of pure MIL-88A and pMIL-88A/PAN NFM, which should be attributed to the fact that the MOF particles are wrapped in the iMIL-88A/PAN and bMIL-88A/PAN NFMs. These results indicate that the in situ electrospun and PVP removing preparation strategy of pMIL-88A/PAN NFM can expose more adsorption sites of MIL-88A for dyes, and more pore structures suitable for dyes adsorption were formed, improving its dye adsorption performance.

2.2.4. Recyclability of the pMIL-88A/PAN NFM

Adsorption–desorption experiments were conducted to evaluate the recyclability of the pMIL-88A/PAN NFM. In the first adsorption test, the pMIL-88A/PAN NFM showed 99–100% removal efficiencies for AR, RB, and AB dyes (Figure 5). Then, the dye-adsorbed NFM was treated with ethanol to destroy the interactions between MOFs and organic dyes and realize the desorption of dyes, so that the NFM can be used for the following recyclability experiments. The desorption efficiency was calculated as the ratio of the mass of NFM desorbed by ethanol to the mass of the initial NFM. The desorption efficiencies of pMIL-88A/PAN NFM adsorbed AR, RB, and AB dyes were as high as 99.9%, indicating that all three dyes could achieve a good adsorption-desorption cycling. After five adsorption–desorption cycles, the removal efficiencies of the pMIL-88A/PAN NFM for AR, RB, and AB dyes remained higher than 80% (Figure 7a). After five adsorption–desorption cycles, the NFM showed good optical shape maintenance (Figure 7b). Moreover, the NFM before RB adsorption and after desorption of RB exhibited similar FT-IR spectra, suggesting the good structural stability of the NFM under long-term application. Overall, the pMIL-88A/PAN NFM exhibited a good recyclability, which enables a long service life and will be beneficial to the recovery and reuse of dyes from wastewater.

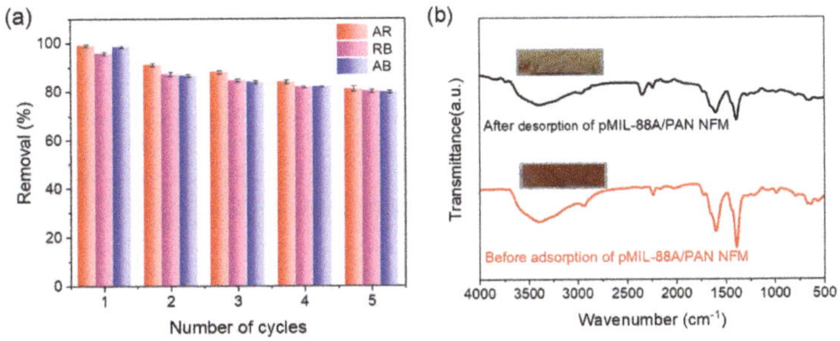

Figure 7. (**a**) Adsorption recyclability of the pMIL-88A/PAN NFM. (**b**) FT−IR spectra of pMIL-88A/PAN NFM before RB adsorption and after desorption of RB; the inset images are the optical photographs of the corresponding NFMs.

3. Materials and Methods

3.1. Materials

PAN (Mw = 200 kDa) was obtained from Shunjie Plastic Technology Co., Ltd. (Najing, China). $FeCl_3$ $6H_2O$ and N,N-dimethylformamide (DMF) were purchased from Tianjin Best Chemical Co., Ltd. (Tianjin, China). PVP (Mw = 130 kDa), fumaric acid (FuA), amaranth red (AR), rhodamine B (RB), and acid blue (AB) were acquired from Energy Chemical (Guangzhou, China). All chemicals were of analytical grade and were used as received without further purification.

3.2. Preparation of NFMs

The NFMs were prepared by electrospinning strategy. Firstly, the PAN/PVP solu-tion (solution A) was prepared by adding 1 g of PAN powder to DMF (PAN:DMF = 1:9 *wt*/*wt*) under constant stirring for 6 h to prepare a uniform and transparent mixture. Thereafter, 1 g of PVP powder was added to the PAN solution and the mixture was stirred continuously for 5 h to prepare the solution A. Next, $FeCl_3$ $6H_2O$ and fumaric acid were added to DMF ($FeCl_3$ $6H_2O$:FuA:DMF = 1.01:2.11:10 *wt*/*wt*), and the mixture was stirred for 3 h to prepare spinning solution B. An electrospinning solution C was prepared by mixing the solution A with the as-prepared solution B.

Subsequently, the solution C was loaded into a 10 mL syringe for electrospinning with an applied voltage of 25 kV, a receiving distance of 15 cm and a flow rate of 0.4 mL h^{-1}. This fiber sample was named in situ MIL-88A/PAN/PVP (iMIL-88A/PAN/PVP) NFM.

The collected iMIL-88A/PAN/PVP NFM was immersed into a mixed aqueous solution containing 50% ethanol at 50 °C for 24 h to remove PVP, followed by rinsing with ethanol and water. Then the NFM was soaked in methanol for 3 d, followed by drying in a vacuum oven at 100 °C overnight to obtain the porous MIL-88A/PAN (pMIL-88A/PAN) NFM.

The PAN/PVP NFM was obtained by electrospinning solution A, and then the pPAN NFM was obtained by removing PVP of PAN/PVP NFM. For comparison, pure MIL-88A particles, in situ MIL-88A/PAN (iMIL-88A/PAN) and blended MIL-88A/PAN (bMIL-88A/PAN) NFMs were also prepared.

3.3. *Characterization*

Scanning electron microscopy (SEM) and elemental mapping measurements were performed on a Sigma500 (ZEISS) scanning electron microscope. Fourier-transform infrared (FT-IR) spectra were recorded on a Nicolet IS 10 (Thermo Fisher, Massachusetts, America) spectrometer. X-ray diffraction (XRD) characterization was performed on a Rigaku SmartLab 9 kW using Cu Kα radiation (λ = 1.5406 Å, 40 kV, and 100 mA). Nitrogen physisorption measurements were conducted to determine the surface areas and pore volumes on a Micromeritics ASAP 2460 apparatus. UV-vis spectra were recorded on an UV-3600 (Shimadzu, Kyoto, Japan) spectrophotometer.

3.4. *Adsorption and Desorption Experiments*

3.4.1. Adsorption Experiments

Kinetic experiments were conducted by adding 20 mg of adsorbent to 100 mL of AR, RB, and AB solutions with initial concentrations of 20 mg L^{-1}. The adsorbent dosage was 0.2 mg mL^{-1}, and 2 mL aliquots were used to determine the concentrations of the dyes. For the adsorption experiments, 10 mg of adsorbent was added to 10 mL of amaranth red, rhodamine B, and acid blue solutions with concentrations in the range of 25–400 mg L^{-1}. For the adsorption–desorption experiments, 20 mg of adsorbent was added to 5 mL of 4 mg mL^{-1} amaranth red, rhodamine B, and acid blue solutions and was allow to react with the dyes for 48 h. The used adsorbents were regenerated by immersing them in ethanol for 4 h to remove the dyes, followed by rinsing with ethanol and water several times and drying under vacuum. The regenerated adsorbents were reused. The dye removal efficiency of each adsorbent sample was determined using the initial adsorption capacity. After five adsorption–desorption cycles the adsorption capacity (q (mg g^{-1})) of each adsorbent for the dyes was calculated as follows:

$$q\ (\text{mg/g}) = \frac{(C_o - C_e)V}{W} \quad (5)$$

where C_o and C_e (mg L^{-1}) are the initial and equilibrium concentrations of amaranth red, rhodamine B, and acid blue in the aqueous solutions, respectively, V (L) is the volume of the dye solution, and W (g) is the mass of the dry adsorbent.

3.4.2. Desorption Experiments

Specifically, the adsorbed nanofiber membranes were immersed in a mixture of ethanol: water = 1:1 for 6 h, followed by washing with ethanol and water several times, and finally the nanofiber membranes were immersed in methanol solution for 12 h, followed by vacuum drying.

4. Conclusions

In summary, we developed a direct electrospinning method to prepare porous MOF/polymer NFMs. MIL-88A MOF particles were formed inside the PAN/PVP fibers in situ during electrospinning, and the pMIL-88A/PAN NFM can be obtained after removing PVP by ethanol and water washing. We demonstrated that the pMIL-88A/PAN NFM achieved high dye removal efficiencies of 99.2, 94.4, and 99.8% for amaranth red, rhodamine B, and acid blue, respectively. Moreover, the pMIL-88A/PAN NFM also presented high adsorption capacities, fast adsorption kinetics, and high cycling stabilities during the processes of dye adsorption and removal. Our results demonstrated that the in situ electrospun porous MOF/polymer NFMs present promising application potential in water treatment for organic dyestuff removal.

Supplementary Materials: The following supporting information can be downloaded at: https://www.mdpi.com/article/10.3390/molecules28020760/s1. The Supporting Information is available free of charge at https://pubs.acs.org/. Preparations of pure MIL-88A powder, in situ MIL-88A/PAN (iMIL-88A/PAN) and blended MIL-88A/PAN (bMIL-88A/PAN) NFMs; Figure S1: SEM images of the PAN/PVP, pPAN; Figure S2: SEM images of the pMIL-88A/PAN NFMs; Figure S3: EDS spectrum of pMIL-88A/PAN NFM; Figure S4: Pore size distributions; Figure S5: Dye removal rates of pPAN NFM, pMIL-88A/PAN NFM, and pure MIL-88A; Figure S6: Adsorption isotherms of MIL-88A powder, iMIL-88A/PAN and bMIL-88A/PAN NFMs; Table S1: Calculated results of the Langmuir and Freundlich models for the adsorption of AR, RB, and AB by MIL-88A powder, iMIL-88A/PAN and bMIL-88A/PAN NFMs. (PDF).

Author Contributions: Supervision, Conceptualization, Writing—review and editing, Project administration, X.L.; Investigation, Data curation, Writing—original draft, H.W.; formal analysis, L.X. and J.J.; Supervision, Writing—review and editing, F.D. and Y.J.; All authors have read and agreed to the published version of the manuscript.

Funding: This work was supported by the National Natural Science Foundation of China (22005224). The research was also supported by Wuyi University (AL2019003).

Institutional Review Board Statement: Not applicable.

Informed Consent Statement: The manuscript was written through contributions of all authors. All authors have given approval to the final version of the manuscript.

Data Availability Statement: Not applicable.

Conflicts of Interest: The authors declare no conflict of interest.

References

1. Rojas, S.; Horcajada, P. Metal–organic frameworks for the removal of emerging organic contaminants in water. *Chem. Rev.* **2020**, *120*, 8378–8415. [CrossRef] [PubMed]
2. Tran, N.H.; Hu, J.; Li, J.; Ong, S.L. Suitability of artificial sweeteners as indicators of raw wastewater contamination in surface water and groundwater. *Water Res.* **2014**, *48*, 443–456. [CrossRef] [PubMed]
3. Buerge, I.J.; Keller, M.; Buser, H.R.; Muller, M.D.; Poiger, T. Saccharin and other artificial sweeteners in soils: Estimated inputs from agriculture and households, degradation, and leaching to groundwater. *Environ. Sci. Technol.* **2011**, *45*, 615–621. [CrossRef] [PubMed]
4. Du, H.; Yue, M.; Huang, X.; Duan, G.; Yang, Z.; Huang, W.; Shen, W.; Yin, X. Preparation, application and enhancement dyeing properties of ZnO nanoparticles in silk fabrics dyed with natural dyes. *Nanomaterials* **2022**, *12*, 3953. [CrossRef] [PubMed]
5. Chen, H.; Huang, M.; Liu, Y.; Meng, L.; Ma, M. Functionalized electrospun nanofiber membranes for water treatment: A review. *Sci. Total. Environ.* **2020**, *739*, 139944. [CrossRef]
6. Baigorria, E.; Galhardi, J.A.; Fraceto, L.F. Trends in polymers networks applied to the removal of aqueous pollutants: A review. *J. Clean. Prod.* **2021**, *295*, 126451. [CrossRef]
7. Dai, X.H.; Fan, H.X.; Yi, C.Y.; Dong, B.; Yuan, S.J. Solvent-free synthesis of a 2D biochar stabilized nanoscale zerovalent iron composite for the oxidative degradation of organic pollutants. *J. Mater. Chem. A* **2019**, *7*, 6849–6858. [CrossRef]
8. Abdi, J.; Abedini, H. MOF-based polymeric nanocomposite beads as an efficient adsorbent for wastewater treatment in batch and continuous systems: Modelling and experiment. *Chem. Eng. J.* **2020**, *400*, 125862. [CrossRef]
9. Zheng, X.; Rehman, S.; Zhang, P. Room temperature synthesis of monolithic MIL-100(Fe) in aqueous solution for energy-efficient removal and recovery of aromatic volatile organic compounds. *J. Hazard. Mater.* **2023**, *442*, 129998. [CrossRef]
10. Yang, W.; Shi, F.; Jiang, W.; Chen, Y.; Zhang, K.; Jian, S.; Jiang, S.; Zhang, C.; Hu, J. Outstanding fluoride removal from aqueous solution by a La-based adsorbent. *RSC Adv.* **2022**, *12*, 30522–30528. [CrossRef]
11. Yang, W.; Wang, Y.; Wang, Q.; Wu, J.; Duan, G.; Xu, W.; Jian, S. Magnetically separable and recyclable Fe_3O_4@ PDA covalent grafted by l-cysteine core-shell nanoparticles toward efficient removal of Pb^{2+}. *Vacuum* **2021**, *189*, 110229. [CrossRef]
12. Gotore, O.; Osamu, N.; Rameshprabu, R.; Arthi, M.; Unpaprom, Y.; Itayama, T. Biochar derived from non-customized matamba fruit shell as an adsorbent for wastewater treatment. *J. Bioresour. Bioprod.* **2022**, *7*, 109–115. [CrossRef]
13. Jjagwe, J.; Olupot, P.W.; Menya, E.; Kalibbala, H.M. Synthesis and application of Granular activated carbon from biomass waste materials for water treatment: A review. *J. Bioresour. Bioprod.* **2021**, *6*, 292–322. [CrossRef]

14. Pintor, A.M.A.; Vilar, V.J.P.; Botelho, C.M.S.; Boaventura, R.A.R. Oil and grease removal from wastewaters: Sorption treatment a an alternative to state-of-the-art technologies. *A critical review. Chem. Eng. J.* **2016**, *297*, 229–255. [CrossRef]
15. Chatzisymeon, E.; Stypas, E.; Bousios, S.; Xekoukoulotakis, N.P. Photocatalytic treatment of black table olive processin wastewater. *J. Hazard. Mater.* **2008**, *154*, 1090–1097. [CrossRef]
16. Dotto, G.L.; McKay, G. Current scenario and challenges in adsorption for water treatment. *J. Environ. Chem. Eng.* **2020**, *8*, 10398 [CrossRef]
17. Ma, X.; Zhao, S.; Tian, Z.; Duan, G.; Pan, H.; Yue, Y.; Li, S.; Jian, S.; Yang, W.; Liu, K.; et al. MOFs meet wood: Reusable magneti hydrophilic composites toward efficient water treatment with super-high dye adsorption capacity at high dye concentratior *Chem. Eng. J.* **2022**, *446*, 136851. [CrossRef]
18. Chen, Y.; Li, S.; Li, X.; Mei, C.; Zheng, J.; E, S.; Duan, G.; Liu, K.; Jiang, S. Liquid transport and real-time dye purification via lotu petiole-inspired long-range-ordered anisotropic cellulose nanofibril aerogels. *ACS Nano* **2021**, *15*, 20666–20677. [CrossRef]
19. Chen, Y.; Hanshe, M.; Sun, Z.; Zhou, Y.; Mei, C.; Duan, G.; Zheng, J.; E, S.; Jiang, S. Lightweight and anisotropic cellulos nanofibril/rectorite composite sponges for efficient dye adsorption and selective separation. *Int. J. Biol. Macromol.* **202** *207*, 130–139. [CrossRef]
20. Ahmadijokani, F.; Molavi, H.; Bahi, A.; Fernandez, R.; Alaee, P.; Wu, S.; Wuttke, S.; Ko, F.; Arjmand, M. Metal-Organic Framework and Electrospinning: A Happy Marriage for Wastewater Treatment. *Adv. Funct. Mater.* **2022**, *25*, 2207723. [CrossRef]
21. Shi, Z.; Tao, Y.; Wu, J.; Zhang, C.; He, H.; Long, L.; Lee, Y.; Li, T.; Zhang, Y.B. Robust metal–triazolate frameworks for CO_2 captur from flue gas. *J. Am. Chem. Soc.* **2022**, *142*, 2750–2754. [CrossRef] [PubMed]
22. Wu, C.; Chou, L.Y.; Long, L.; Si, X.; Lo, W.S.; Tsung, C.K.; Li, T. Structural control of uniform MOF-74 microcrystals for the stud of adsorption kinetics. *ACS Appl. Mater. Inter.* **2019**, *11*, 35820–35826. [CrossRef] [PubMed]
23. Cui, Y.; Li, B.; He, H.; Zhou, W.; Chen, B.; Qian, G. Metal–organic frameworks as platforms for functional materials. *Acc. Chen Res.* **2016**, *49*, 483–493. [CrossRef] [PubMed]
24. Deng, Y.; Wu, Y.; Chen, G.; Zheng, X.; Dai, M.; Peng, C. Metal-organic framework membranes: Recent development in th synthesis strategies and their application in oil-water separation. *Chem. Eng. J.* **2021**, *405*, 127004. [CrossRef]
25. Fan, L.; Yu, Q.; Chen, J.; Khan, U.; Wang, X.; Gao, J. Achievements and Perspectives in Metal–Organic Framework-Based Material for Photocatalytic Nitrogen Reduction. *Catalysts* **2022**, *12*, 1005. [CrossRef]
26. Wang, X.; Yang, X.; Chen, C.; Li, H.; Huang, Y.; Cao, R. Graphene Quantum Dots Supported on Fe-based Metal-Organi Frameworks for Efficient Photocatalytic CO_2 Reduction. *Acta. Chim. Sin.* **2022**, *80*, 22–28. [CrossRef]
27. Li, T.; Liu, L.; Zhang, Z.; Han, Z. Preparation of nanofibrous metal-organic framework filter for rapid adsorption and selectiv separation of cationic dye from aqueous solution. *Sep. Purif. Technol.* **2020**, *237*, 116360. [CrossRef]
28. Seo, P.W.; Khan, N.A.; Jhung, S.H. Removal of nitroimidazole antibiotics from water by adsorption over metal–organic framework modified with urea or melamine. *Chem. Eng. J.* **2017**, *315*, 92–100. [CrossRef]
29. Kalaj, M.; Bentz, K.C.; Ayala, S., Jr.; Palomba, J.M.; Barcus, K.S.; Katayama, Y.; Cohen, S.M. MOF-polymer hybrid materials: Fror simple composites to tailored architectures. *Chem. Rev.* **2020**, *120*, 8267–8302. [CrossRef]
30. Xue, J.; Wu, T.; Dai, Y.; Xia, Y. Electrospinning and electrospun nanofibers: Methods, materials, and applications. *Chem. Rev.* **201** *119*, 5298–5415. [CrossRef]
31. Jian, S.; Chen, Y.; Shi, F.; Liu, Y.; Jiang, W.; Hu, J.; Han, X.; Jiang, S.; Yang, W. Template-Free Synthesis of Magnetic La-Mn-F Tri-Metal Oxide Nanofibers for Efficient Fluoride Remediation: Kinetics, Isotherms, Thermodynamics and Reusability. *Polymer* **2022**, *14*, 5417. [CrossRef]
32. Venkatesan, M.; Veeramuthu, L.; Liang, F.C.; Chen, W.C.; Cho, C.J.; Chen, C.W.; Chen, J.Y.; Yan, Y.; Chang, S.H.; Kuo, C.K Evolution of electrospun nanofibers fluorescent and colorimetric sensors for environmental toxicants, pH, temperature, anc cancer cells–A review with insights on applications. *Chem. Eng. J.* **2020**, *397*, 125431. [CrossRef]
33. Zhao, R.; Shi, X.; Ma, T.; Rong, H.; Wang, Z.; Cui, F.; Zhu, G.; Wang, C. Constructing mesoporous adsorption channels anc MOF–polymer interfaces in electrospun composite fibers for effective removal of emerging organic contaminants. *ACS App Mater. Inter.* **2020**, *13*, 755–764. [CrossRef]
34. Zhao, R.; Tian, Y.; Li, S.; Ma, T.; Lei, H.; Zhu, G. An electrospun fiber based metal–organic framework composite membran for fast, continuous, and simultaneous removal of insoluble and soluble contaminants from water. *J. Mater. Chem. A* **2019** *7*, 22559–22570. [CrossRef]
35. Jian, S.; Cheng, Y.; Ma, X.; Guo, H.; Hu, J.; Zhang, K.; Jiang, S.; Yang, W.; Duan, G. Excellent fluoride removal performance by electrospun La–Mn bimetal oxide nanofibers. *New J. Chem.* **2022**, *46*, 490–497. [CrossRef]
36. Jian, S.; Tian, Z.; Zhang, K.; Duan, G.; Yang, W.; Jiang, S. Hydrothermal synthesis of Ce-doped ZnO heterojunction supported or carbon nanofibers with high visible light photocatalytic activity. *Chem. Res. Chin. Univ.* **2021**, *37*, 565–570. [CrossRef]
37. Mahmoodi, N.M.; Oveisi, W.; Taghizadeh, A.; Taghizadeh, M. Synthesis of pearl necklace-like ZIF-8@chitosan/PVA nanofibe with synergistic effect for recycling aqueous dye removal. *Carbohyd. Polym.* **2020**, *227*, 115364. [CrossRef]
38. Dou, Y.; Zhang, W.; Kaiser, A. Electrospinning of metal–organic frameworks for energy and environmental applications. *Adv. Sc* **2020**, *7*, 1902590. [CrossRef]
39. Lu, A.X.; McEntee, M.; Browe, M.A.; Hall, M.G.; DeCoste, J.B.; Peterson, G.W. MOFabric: Electrospun nanofiber mats fror PVDF/UiO-66-NH2 for chemical protection and decontamination. *ACS Appl. Mater. Inter.* **2017**, *9*, 13632–13636. [CrossRef]

40. Xie, A.; Cui, J.; Yang, J.; Chen, Y.; Lang, J.; Li, C.; Yan, Y.; Dai, J. Graphene oxide/Fe (III)-based metal-organic framework membrane for enhanced water purification based on synergistic separation and photo-Fenton processes. *Appl. Catal. B-Environ.* **2020**, *264*, 118548. [CrossRef]
41. Leus, K.; Krishnaraj, C.; Verhoeven, L.; Cremers, V.; Dendooven, J.; Ramachandran, R.K.; Dubruel, P.; Van der Voort, P. Catalytic carpets: Pt@ MIL-101@ electrospun PCL, a surprisingly active and robust hydrogenation catalyst. *J. Catal.* **2018**, *360*, 81–88. [CrossRef]
42. Chen, C.; Zhang, W.; Zhu, H.; Li, B.G.; Lu, Y.; Zhu, S. Fabrication of metal-organic framework-based nanofibrous separator via one-pot electrospinning strategy. *Nano Res.* **2021**, *14*, 1465–1470. [CrossRef]
43. Xue, B.; Du, L.; Jin, J.; Meng, H.; Mi, J. In situ growth of MIL-88A into polyacrylate and its application in highly efficient photocatalytic degradation of organic pollutants in water. *Appl. Surf. Sci.* **2021**, *564*, 150404. [CrossRef]
44. Viswanathan, V.P.; Mathew, S.V.; Dubal, D.P.; Adarsh, N.N.; Mathew, S. Exploring the Effect of Morphologies of Fe (III) Metal-Organic Framework MIL-88A (Fe) on the Photocatalytic Degradation of Rhodamine B. *ChemistrySelect* **2020**, *5*, 7534–7542. [CrossRef]
45. Wang, L.; Zhang, Y.; Li, X.; Xie, Y.; He, J.; Yu, J.; Song, Y. The MIL-88A-derived Fe_3O_4-carbon hierarchical nanocomposites for electrochemical sensing. *Sci. Rep.* **2015**, *5*, 14341. [CrossRef]
46. Liu, W.; Zhou, J.; Ding, L.; Yang, Y.; Zhang, T. MIL-88/PVB nanofiber as recyclable heterogeneous catalyst for photocatalytic and Fenton process under visible light irradiation. *Chem. Phys. Lett.* **2020**, *749*, 137431. [CrossRef]
47. He, J.; Wang, W.; Sun, F.; Shi, W.; Qi, D.; Wang, K.; Shi, R.; Cui, F.; Wang, C.; Chen, X. Highly efficient phosphate scavenger based on well-dispersed La $(OH)_3$ nanorods in polyacrylonitrile nanofibers for nutrient-starvation antibacteria. *ACS Nano* **2015**, *9*, 9292–9302. [CrossRef]
48. Ninwiwek, N.; Hongsawat, P.; Punyapalakul, P.; Prarat, P. Removal of the antibiotic sulfamethoxazole from environmental water by mesoporous silica-magnetic graphene oxide nanocomposite technology: Adsorption characteristics, coadsorption and uptake mechanism. *Coll. Surf. A* **2019**, *580*, 123716. [CrossRef]

molecules

MDPI

Article

Valorization of CO_2 through the Synthesis of Cyclic Carbonates Catalyzed by ZIFs

José J. Delgado-Marín [1], Iris Martín-García [2], David Villalgordo-Hernández [1], Francisco Alonso [2], Enrique V. Ramos-Fernández [1,*] and Javier Narciso [1,3]

1 Instituto de Materiales, Departamento de Química Inorgánica, Facultad de Ciencias, Universidad de Alicante, Apdo. 99, 03080 Alicante, Spain
2 Instituto de Síntesis Orgánica, Departamento de Química Orgánica, Facultad de Ciencias, Universidad de Alicante, Apdo. 99, 03080 Alicante, Spain
3 Instituto de Investigación Sanitaria Biomédica de Alicante (ISABIAL), 03690 Alicante, Spain
* Correspondence: enrique.ramos@ua.es

Abstract: One way to exploit CO_2 is to use it as a feedstock for the production of cyclic carbonates via its reaction with organic epoxides. As far as we know, there is still no heterogeneous catalyst that accelerates the reaction in a selective, efficient and industrially usable way. Cobalt and zinc-based zeolitic imidazole frameworks (ZIFs) have been explored as heterogeneous catalysts for this reaction. In particular, we have prepared ZIF-8 and ZIF-67 catalysts, which have been modified by partial replacement of 2-methylimidazole by 1,2,4-triazole, in order to introduce uncoordinated nitrogen groups with the metal. The catalysts have shown very good catalytic performance, within the best of the heterogeneous catalysts tested in the cycloaddition of CO_2 with epichlorohydrin. The catalytic activity is due ultimately to defects on the outer surface of the crystal, and varies in the order of ZIF-67-m > ZIF-67 > ZiF-8-m = ZIF-8. Notably, reactions take place under mild reaction conditions and without the use of co-catalysts.

Keywords: carbon dioxide; cyclic carbonates; epichlorohydrin; epoxides; heterogeneous catalysis; ZIFs

1. Introduction

Nowadays, one of the main environmental challenges is the reduction of the greenhouse effect, caused by a reduction in the generation of certain gases (especially CO_2). The easiest way to reach this goal is to decrease the use of fossil fuels by substitution with renewable energies, the use of more efficient engines, and hybrid or electric cars. Norway is a paradigmatic country where electricity is practically obtained from renewable energies. Different strategies have been proposed for CO_2 mitigation, perhaps the most effective so far being to utilize CO_2 for the preparation of cyclic carbonates (and also at an industrial level) through the cycloaddition of CO_2 with epoxides [1]. Obviously, the production of cyclic carbonates and polycarbonates is very low compared to the global release of CO_2 to the atmosphere, but it is an avenue for sequestering CO_2 that should be explored (since apart from the current uses of polycarbonates, these could replace some of the plastics we use today). The two most widely used polycarbonates are based on ethylene oxide and propylene oxide, both of which are obtained from the petroleum industry (which is not a renewable source). However, efforts are being made to find other routes for the production of these precursors, such as those based on bioalcohols derived from biomass [2]. Limonene oxide has also been described as an alternative epoxide, whose precursor monoterpene is obtained from the citrus peel used in the production of juices [3]. Perhaps the most interesting alternative is the use of glycerol as a precursor of glycidyl alcohol, which is a by-product in the synthesis of biodiesel. However, to date, the results obtained from this approach have not been very promising [4–6]. However, epichlorohydrin can be obtained from glycerol, which is one of the most widely used epoxides to synthesize cyclic carbonates, at least at an academic level [7–10]. These types of reactions are catalyzed both under homogeneous and heterogeneous conditions. In homogeneous catalysis, the most used catalysts are, for example, quaternary ammonium salt halides (tetrabutylammonium halides, TBAX) and imidazolium halides (with regards to Lewis-base homogeneous catalysis) [11,12]. In the case of acid catalysis, metal porphyrins have been widely used [13,14]. The main problem of homogeneous catalysis is the need to separate the catalyst from the products, something that is usually carried out by distillation, which is a large energy-consumer purification technique. In heterogeneous catalysis, one of the most used routes is the immobilization of an active phase (Lewis base) on a support, such as MCM41 [15,16] or SBA-15 [17,18]. Concerning acidic sites, Cr-MIL-101 [19,20] is one of the most efficient catalysts, where the metal loses two coordination water molecules when it is activated, forming what is called the open metal site (Lewis acid character). However, it is known that both ZIF-8 and ZIF-67 are also catalytically active [21,22], which in principle should not be. These three latter catalysts are MOFs (metal organic frameworks [23,24]), which are basically metallic nodes or clusters linked through organic ligands, which generate a three-dimensional structure with a high porosity that can exceed 4000 m^2/g in some cases. The advantage of ZIFs (zeolitic imidazolate framework), especially ZIF-8 and ZIF-67, over most MOFs is their high thermal (>400 °C, air) and water vapor stability. In addition, at present, there are already synthetic routes to MOFs that are found within the so-called 'Green Chemistry', since their synthesis is carried out in an aqueous medium at moderate pH. ZIF-8 and ZIF-67 are isostructural, the difference between them is that Zn^{2+} is the cation in the first case and Co^{2+} in the second case. Unlike Cr-MIL-101, ZIF-8 and ZIF-67 are catalytically active since ZIFs have some defects (that is to say, the metal node can have coordinatively unsaturated sites displaying Lewis acid behavior) [25]. As the coordination is carried out with the nitrogen atoms of 2-methylimidazole (linker), this implies that a possible adjacent basic center could be generated. One of the possible strategies to improve the efficiency of these ZIFs could be to search for synthetic routes that promote structural defects, or as a more elegant approach, the partial substitution of 2-methylimidazole for a triazole to significantly increase the number of basic sites.

In the present investigation, four catalysts have been synthesized (ZIF-8, ZIF-67, ZIF-8-m and ZIF-67-m) where 'm' indicates that there has been a partial substitution of 2-methylimidazole for a 1,2,4-triazole, maintaining the ZIF structure. Their catalytic activity has been compared in the reaction of CO_2 with epichlorohydrin to give the corresponding cyclic carbonate (Scheme 1). The effect of the temperature, pressure, reaction time, and the incorporation of TBAI as a cocatalyst have also been studied.

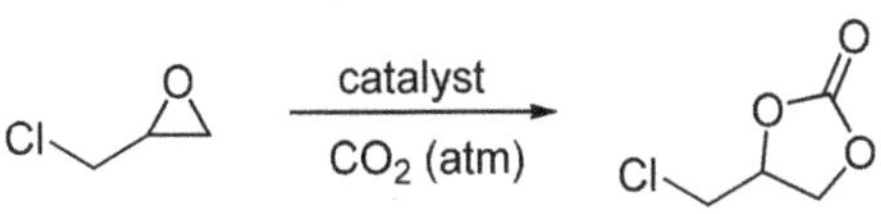

Scheme 1. Reaction of fixation of CO_2 in epichlorohydrin to give 3-chloropropene carbonate.

2. Results and Discussion

Four catalysts have been prepared, which are in principle isostructural, something that can be clearly seen in the diffractograms shown in Figure 1. The diffraction patterns of ZIF-8 and ZIF-67 are exactly the same as those previously published. However, the catalysts modified with 1,2,4-triazole present certain differences, especially ZIF-67, where the peaks appear shifted and there is a variation in the relative intensity. This indicates that the exchange of the ligands produces a change in the unit cell and a small loss of crystallinity. It is important to remember that Zubieta and co-workers have carried out an exhaustive study on the synthesis of MOFs [26,27] with bimetallic cations and 1,2,4-triazole, and in no case they obtained an isostructural compound with ZIF-8. Thus, a topological change could be induced if the degree of exchange is very high. In the case of ZIF-8, a shift to larger angles is mainly observed, indicating that the structure is the same but, as expected, the lattice parameters change since the ligand is slightly smaller.

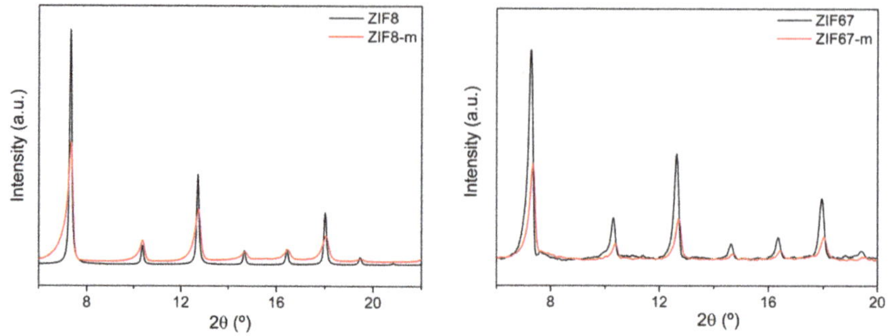

Figure 1. X-ray diffraction patterns of the four catalysts.

Figure 2 and Table 1 show the results of adsorption of the four catalysts. The surface area does experience a great change, especially in ZIF-67, as the observed isotherm is type I in all cases, albeit ZIF-67-m seems to present a slight mesoporosity. But the most remarkable thing is that the typical step at low relative pressures appears for all of these catalysts, which is comparable to the fingerprint of these materials. In principle, it would be expected that with such a drastic reduction in the area of ZIF-8-m and ZIF-67-m (60 and 50% respectively), the adsorption capacity of the catalysts should be less efficient, even though the incorporation an extra N atom into the structure should favor the adsorption of CO_2. The CO_2 adsorption isotherms are very interesting: it can be seen that there is no difference in the type of adsorption for ZIF-8 and modified ZIF-8, both cases being linear, but the maximum adsorption capacity only decreases a 10%, while the decrease in the surface area is a 40%.

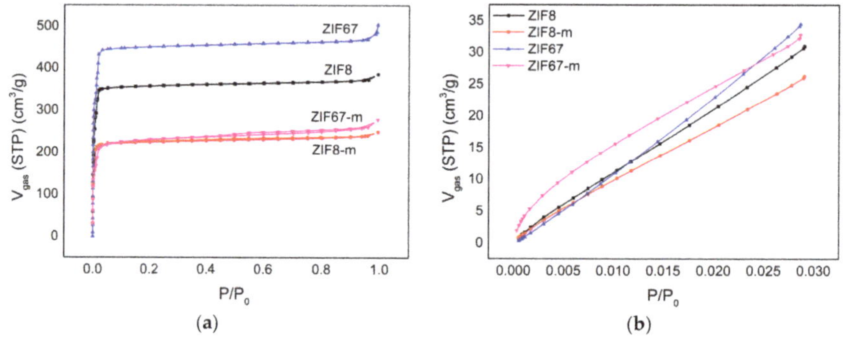

(**a**) (**b**)

Figure 2. Adsorption isotherm of the four catalysts. (**a**) N_2 at −196 °C; (**b**) CO_2 at 0 °C.

Table 1. Textural properties of the four catalysts obtained from N_2 adsorption.

Sample	BET (m^2/g)	Micropore Volume (cm^3/g)	Pore Volume (cm^3/g)
ZIF-8	1551	0.55	0.58
ZIF-8-m	942	0.35	0.37
ZIF-67	1833	0.70	0.72
ZIF-67-m	914	0.36	0.40

Much more notable is the case of ZIF-67, where the reduction in the CO_2 adsorption capacity is only 10%, while that of the surface area is almost 60%, as mentioned above, and now it is no longer linear but rather shows a greater affinity for CO_2 than in the case of the ZIF-67. The aforementioned shows that, in this case, there must be a change of greater scope than just the replacement of the linkers.

A semi-quantitative determination of the degree of exchange can be carried out with the TGA-MS technique. The protocol used was as follows. First, the catalyst was heated to 900 °C in an Ar atmosphere. Then, the sample was cooled to 50 °C and the process was repeated, but this time in a 4:1 Ar/O_2 mixture. Air was not used to avoid overlapping of N_2 signals with those of CO. In the first stage, a carbon material with a high nitrogen content and the metal [11] was obtained; in the second stage, the carbon phase was eliminated and the metal remained either as ZnO or as Co_3O_4, allowing us to approximately quantify how much linker was exchanged. In the case of ZIF-67 (Figure A2) and ZIF-8, a single very well-defined weight loss was observed around 600 °C, and the final mass of oxide allowed us to identify that everything was correct (99% in both cases). This small difference can be attributed to an error in the quantification, such as the presence in the sample of a small amount of solvent, some defects, etc. Figure 3 shows the TGA-MS of the modified catalysts. First of all, it should be noted that the TGA curves are more complex, basically observing two zones, one at low temperature (300–400 °C) and another more irregular above 600 °C. This indicates that the material is much less stable. In addition, it is observed by MS that both linkers come out mainly at much lower temperatures than that of the unmodified ZIFs, and that the weight loss is greater in the first stage (Ar), followed by a second weight loss that can be attributed to the formation of a carbon material.

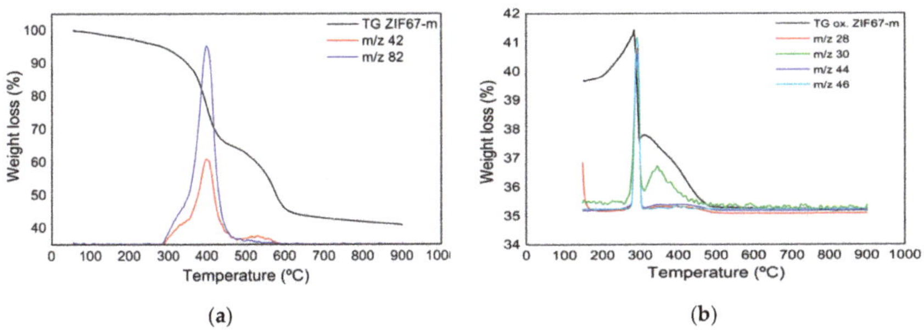

Figure 3. TGA-MS of sample ZIF-67-m. (**a**) In Ar, where *m/z* 42 is related with the 1,2,4-triazole and *m/z* 82 is related with 2-methylimidazole; (**b**) in 4:1 $Ar{:}O_2$, where *m/z* 28 is CO, *m/z* 30 is NO, *m/z* 44 is CO_2, and *m/z* 46 is NO_2.

In the second stage, we observed the combustion of the carbon material and the oxidation of the metal. It is also notable that the C/N ratio decreases quite a bit based on the ratio of the CO_2/NO_2 areas, indicating that the formed carbon material contains a higher N content, what is very interesting. Now the content of ZnO and Co_3O_4 is different. If we assume that no other phase is being formed, we can deduce that the degree of exchange is 25% in the case of ZIF-8 and 30% in the case of ZIF-67.

Figure 4 shows the XPS spectrum of the four catalysts developed; only the N signal is shown since, in principle, it is the most relevant, while the metal spectra are shown in the Appendix A (Figure A3). Before analyzing the spectra in detail, it must be noted that the nitrogen-to-metal ratio increases with the modified catalysts (as expected). In principle, the N atom not bound to N should give a different signal from that bound. It is clearly observed that there is a use of the non-coordinated N signal. The presence of the said N in the unchanged catalyst is indicative that there are defects in the coordination. In the case of the ZIF-67-m, the largest signal may be indicative that there are a greater number of defects or that a new phase is being formed where the coordination may be changing.

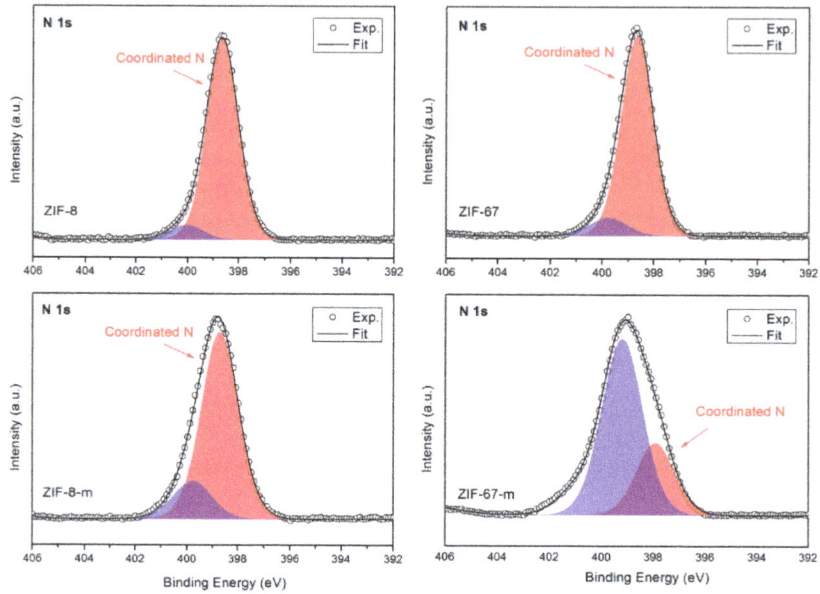

Figure 4. XPS spectra of the four catalysts.

Figure 5 shows two electron micrographs of ZIF-67, where it can be seen that it has a very regular distribution where crystals with a truncated octahedron shape predominate, with an average size of about 400 nm. In the exchange process, no modification has been observed in the microstructure or in the surface of the faces of the monocrystals.

Figure 5. Secondary electron micrographs of ZIF-67.

In order to obtain the conversion of epichlorohydrin to 3-chloropropene carbonate, the reaction products were analyzed by ^{1}H NMR. Figure A1 shows the ^{1}H-NMR spectra in great detail, where it can be clearly seen that the only compounds appearing in the reaction crude are the starting epichlorohydrin and its derived carbonate. Figure 6 shows the ^{1}H NMR spectrum of epichlorohydrin and a typical spectrum of one of the reaction-crude aliquots after filtration. The fact that the spectra are very clean makes the quantification of both compounds simple. Given that no other products have been detected in any of the experiments carried out, the selectivity achieved is 100%.

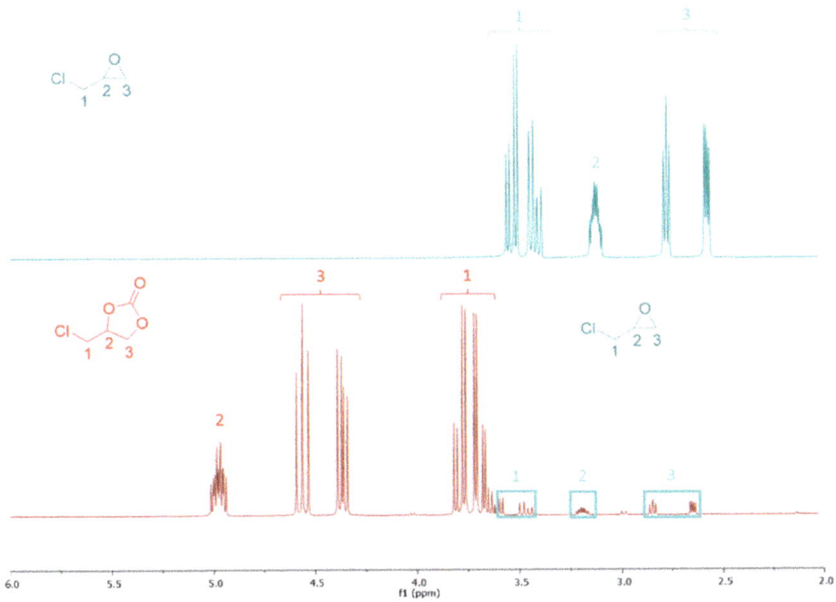

Figure 6. ^{1}H-NMR spectra of epichlorohydrin (blue) and a reaction crude (red).

As indicated in the introduction, tetrabutylammonium halide salts are good homogeneous catalysts for the title reaction and, therefore, they can be also used as cocatalysts [28]. First of all, its effect and its obligatory presence in the reaction have been verified. As it can be seen in Figure 7a, the presence of TBAI improves the catalytic results of ZIF-8, but ZIF-8 itself already shows good catalytic behavior; that is why the use of TBAI was not considered necessary.

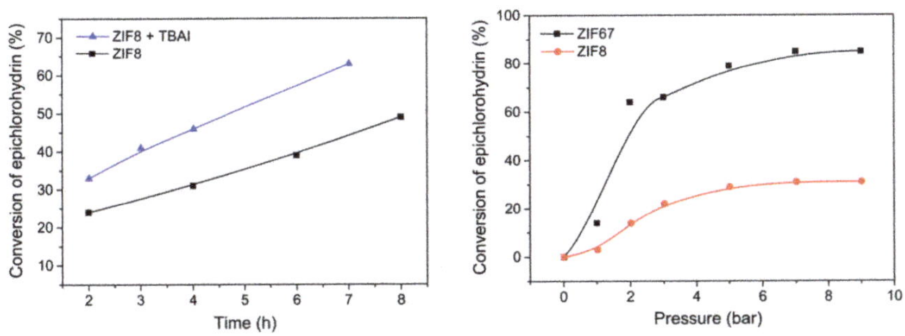

(a) (b)

Figure 7. (**a**) Study of the effect of co-catalyst in the reaction of CO_2 with epichlorohydrin using ZIF-8 as catalyst. Conditions: 97.8 mol.% of epichlorohydrin, 0.6 mol.% of ZIF-8, 1.4 mol.% of TBAI, 60 °C, 7 bar CO_2. (**b**) Effect of the pressure in the catalytic reaction of cycloaddition of CO_2 with two different catalysts (ZIF-8 and ZIF-67). Conditions: 99.4 mol.% of epichlorohydrin, 0.6 mol.% of ZIF catalysts, 120 °C, 4 h. The conversion of epichlorohydrin was based on ^{1}H-NMR results (3-chloropropene carbonate was the only product).

The three main variables studied in the reaction are CO_2 pressure, time, and temperature. First, the effect of pressure on the reaction was analyzed. The data obtained are shown in Figure 7b. As expected, there is a rapid increase in the conversion with pressure, which becomes steady from 7 bars. Some authors [29] obtained similar results, and commented that the conversion decreased drastically if the pressure was greater than 20 bars; we believe that this behavior maybe a consequence of the collapse of the MOF structure.

The effect of the temperature is shown for ZIF-8-m, where a linear increase in conversion with temperature can be seen, although it is usually exponential, which indicates that there are other factors controlling the reaction rate.

Once the pressure and temperature were established, we analyzed the effect of time (Figure 8b), where it is clearly observed that ZIF-67 is a better catalyst than ZIF-8, and that the modified catalyst in the case of ZIF-67 does experience an improvement, while in the case of the ZIF-8 it is not appreciable. Although, apparently, the improvement is not substantial, in reality it is very remarkable since, if we take into account the decrease in the volume of the micropores, the activity of the ZIF-67-m catalyst per unit area is 268% more effective. This actually occurs since the extra N atom in the triazole is even more effective than the other two nitrogen atoms coordinated to the metal, according to the mechanism we have proposed.

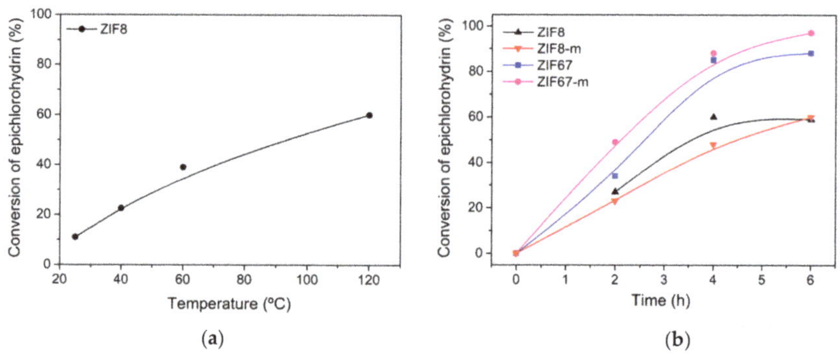

Figure 8. (**a**) Effect of the temperature in the catalytic cycloaddition of CO_2 with epichlorohydrin using ZIF-8. Conditions: 99.4 mol.% of epichlorohydrin, 0.6 mol.% of ZIF catalysts, 7 bar CO_2, 6 h. (**b**) Catalytic reaction of cycloaddition of CO_2 with four different catalysts (ZIF-8, ZIF-8-m, ZIF-67 and ZIF-67-m). Conditions: 99.4 mol.% of epichlorohydrin, 0.6 mol.% of ZIF catalysts, 7 bar CO_2, 120 °C. The conversion of epichlorohydrin was based on ^{1}H-NMR results (3-chloropropene carbonate was the only product).

Concerning the reaction mechanism, different reports describe how ZIF-67 and ZIF-8 materials can activate the epoxide and CO_2 towards the cycloaddition reaction (Figure 9a) [7,21]. It is generally assumed that the acidic and basic sites on the external surface and/or structural defects can account for the catalysis [10,30]. The atmospheric moisture and CO_2 can contribute to the formation of superficial OH and NH groups, together with hydrogenocarbonates [31]. Although these groups might activate the epoxide by hydrogen bonding [10], the presence of low-coordinated Co(II,III) ions is expected to exert a stronger activating effect on the epoxide as Lewis acids. These kind of defects can leave free N atoms in the 2-imidazolate ligands, which have been suggested to activate CO_2 by nucleophilic attack on the carbon atom. Nucleophilic attack of the resulting species on the Co-activated epoxide, followed by intramolecular cyclization, would give the corresponding cyclic carbonate (Figure 9(b1)). In our opinion, the possibility of superficial OH activating CO_2 must not be ruled out [10] (Figure 9(b2)). Indeed, coordination of the Lewis-acidic Co ion to the pyrrolic nitrogen of 2-methylimidazole could have an electron withdrawing effect on the heterocyclic ring, decreasing its nucleophilic character and, consequently, the activation power of the free N atom on CO_2, unless the 2-methyl and the metal-ligand backbonding could compensate for this effect [32]. An additional decrease in the reactivity of the imidazolate unit due to the steric effect of the 2-methyl group must not be disregarded.

The possibility of a coordinated N atom to the metal (N-Co) activating CO_2 must be totally discarded due to its depleted nucleophilic character and steric constraint.

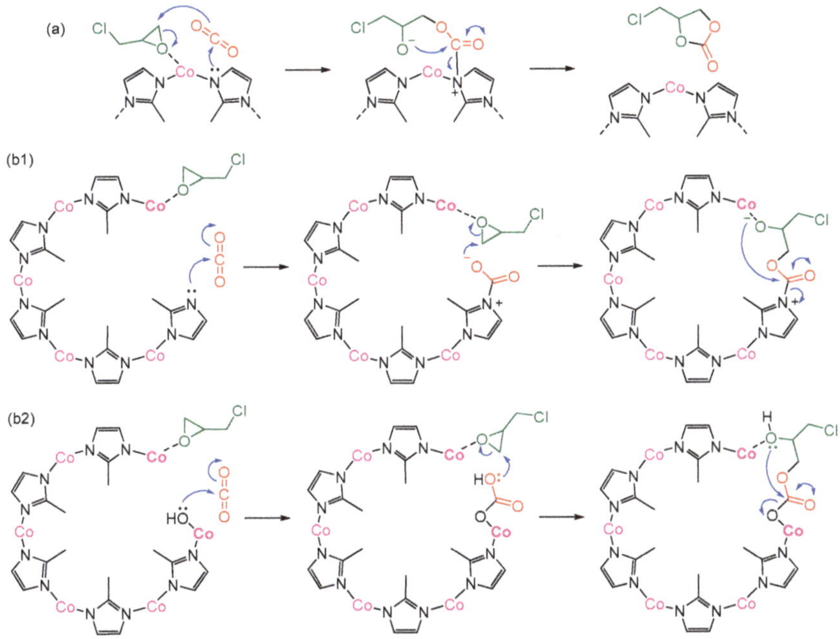

Figure 9. (**a**) Classical reaction mechanism for the cycloaddition of epichlorohydrin and CO_2 catalyzed by ZIF-67. (**b**) Proposed reaction mechanism with ZIF-67 involving structural defects. (**b1**) CO_2 activation by a partially coordinated 2-methylimidazole unit. (**b2**) CO_2 activation by a superficial hydroxyl group.

As regards 1,2,4-triazole, different theoretical and experimental studies support that the tautomeric form A predominates or is the exclusive one (Figure 10a) [33,34]. It shows a decreased basicity (pKa 10.3) with respect to that of 2-methylimidazole (pKa 14.4) and it is deactivated against electrophilic attack, thus resembling the electronic character of pyridine [35]. We believe that the reduced basicity of 1,2,4-triazole favors the formation of more defects when introduced into the structure of the ZIF (Figure 10b), increasing the proportion of low-coordinated Co(II,III) ions and improving the performance of the catalyst (Figure 10b). The lack of nucleophilic character makes the activation of CO_2 by a 1,2,4-triazolate unit very improbable. However, an activation similar to that in Figure 9b could take place, where two 2-methylimidazolate units could participate, one activating the epoxide and the other one CO_2 (the latter through Co-OH species). However, if two vacants are available on Co, activation of both CO_2 and the epoxide on the same site would be also feasible (Figure 10c). This proposal is particularly interesting if we take into account that the same Co site would put into spatial proximity both components, the epoxide and CO_2.

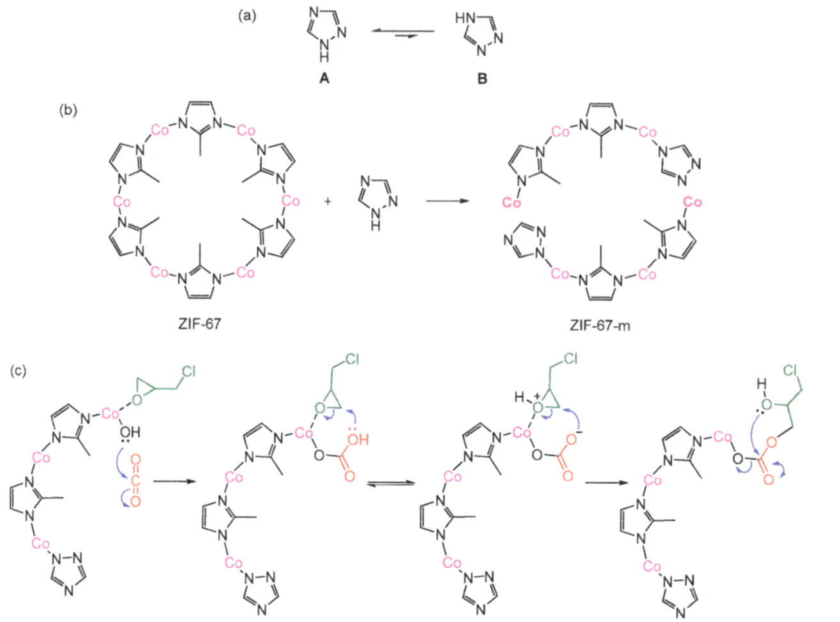

Figure 10. (**a**) Tautomeric forms of 1,2,4-triazole. (**b**) Proposed structural modification of ZIF-67 with 1,2,4-triazole. (**c**) Proposed reaction mechanism for the ZIF-67-m-catalyzed cycloaddition of epichlorohydrin and CO_2.

3. Materials and Methods

The synthesis of catalysts followed a procedure as reported elsewhere [36,37], in which 144 mmol of 2-methylimidazole were dissolved in 50 mL of deionized water and 12 mmol of cobalt(II) or zinc(II) acetate were dissolved in 25 mL of deionized water. Then, both solutions were blended and stirred vigorously during 5 min and kept during 72 h at room temperature; any of the MOF precipitates was separated by centrifugation and washed thrice with MeOH. Finally, the MOF powder was dried for 24 h at 60 °C in a conventional oven. All chemicals have been provided by Aldrich (analytical grade) and has been used without further purification. Once the MOF was prepared, the partial transformation of ZIF-8 or ZIF-67 into ZIF-8-m or ZIF-67-m has been accomplished following the next procedure: 200 mg of ZIF were blended with 200 mg of 1,2,4-triazole; the blend was placed in a vial, the vial was then flushed with N_2 and sealed with a silicone cap. The sealed vial was placed in a pre-heated oven at 130 °C and kept in the oven for 8 h. After that, the vial was broken and a solid was obtained. The solid was composed of ZIF-8-m or ZIF-67-m, together with some impurities (i.e., unreacted linker). Purification was performed by Soxhlet extraction, using acetone as a solvent for 24 h.

The porosity of the samples was characterized by means of nitrogen adsorption-desorption isotherms and CO_2 adsorption isotherms. Samples were outgassed at 150 °C for 4 h prior to the adsorption measurements. The nitrogen adsorption-desorption isotherms were measured at –196 °C in a Quadrawin (Quantachrome) device. S_{BET} was determined from the N_2 adsorption branch. In all cases, the number of points used to apply the BET equation was higher than 5, and the value of c was always positive. The CO_2 adsorption isotherms were measured at 0 °C in the same device. V_{micro} was estimated by the Dubinin-Raduskevich method, with the objective to determine whether any diffusional restrictions in the adsorption took place.

Crystallographic phases were identified by powder X-ray diffraction (PXRD), recorded on a Brucker D8-Advanced diffractometer with a Goebel mirror and a Kristalloflex K 760-

80F X-ray generation system (K-alfa, λ = 1.54 Å), fitted with a Cu cathode and a Ni filter. Spectra were registered between 3° and 40° with a step of 0.05° and a step time of 3 s.

XPS measurements were acquired in a VG-Microtech Multilab device (VG-Microtech, UK) with a Mg-Kα (Hv: 1253.6 eV) radiation source, pass energy of 50 eV and background pressure of 5×10^{-7} Pa. A careful deconvolution of the spectra was made; the areas of the peaks were estimated by calculating the integral of each peak after subtracting a Shirley background and fitting the experimental peak to a combination of Lorentzian/Gaussian lines of 30/70 proportions.

The morphology of the samples was studied by Field Emission Scanning Electron Microscopy with X-ray microanalysis (FESEM–EDS) (ZEISS-Merlin VP Compact, BRUKER-Quantax 400) in both, Backscattered Electron (BSE) and Secondary Electron (SE) modes.

The characterization of the samples by means of TG-DTA-MS was carried out in a TGA/SDTA851e/LF/1600 apparatus from Mettler Toledo, equipped with the Thermostar GSD301T Pfeiffer mass spectrometer. The TG experiments were carried out in the dynamic atmosphere of Ar (100 cm^3/min), with a heating rate of 10 °C/min, while scanning masses up to 200 amu.

Reactions were carried out in a stainless-steel HPM-Vivor autoclave (Premex Solutions Gmbh); volume = 60 mL; P_{max} = 300 bar; T_{max} = 200 °C. The epoxide (epichlorohydrin, 18.0 mmol), catalyst (0.108 mmol) and, in some experiments, quaternary ammonium salt (TBAI, 0.288 mmol) were added to the reactor. The reactor was purged three times followed by the introduction of CO_2 gas (1–9 bar,). The reaction mixture was stirred at 25–120 °C for 1–12 h. The resulting mixture was cooled to room temperature and collected in a vial to do the analysis by ^{1}H-NMR.

^{1}H-NMR spectra were recorded on Bruker Avance 300 and 400 spectrometers (300 and 400 MHz, respectively); chemical shifts are shown in parts per million (δ) and coupling constants (*J*) in Hertz (Hz). The ^{1}H-NMR conversions were determined from the reaction crudes using mesitylene as an internal standard or the signal of the limiting reagent. Samples (5–10 mg) were taken from the reaction crude after filtration through a pad containing layers of Celite and $MgSO_4$, and were diluted with $CDCl_3$ (0.5 mL) as solvent. The ^{1}H-NMR spectra are shown in the Appendix A (Figure A1).

4. Conclusions

Four catalysts have been synthesized with the same structure based on ZIF-8, where the metal Zn has been changed into Co to generate ZIF-67. The linker in ZIF-8 and ZIF-67, 2-methylimidazole, has been partially exchanged for 1,2,4-triazole, giving rise to ZIF-8-m and ZIF-67-m. The degree of exchange is approximately 25% for both ZIFs. The exchanged catalysts (ZIF-8-m, ZIF-67-m) show a BET area decrease between 40–60% while the CO_2 adsorption capacity practically does not vary. An important fact is that Co^{2+} is partially oxidized to Co^{3+}.

In the catalytic process, it has been found that the four catalysts are very active in obtaining the cyclic carbonate derived from CO_2 and epichlorohydrin. It is noteworthy that mild conditions have been applied and the use of cocatalysts has not been necessary. The catalytic activity varies in the following order: ZIF-67-m > ZIF-67 > ZIF-8-m = ZIF-8. The higher catalytic activity of the modified catalysts has been mechanistically rationalized in terms of structural defects.

Author Contributions: J.J.D.-M., D.V.-H., experiments; I.M.-G., F.A. methodology, validation, writing-review-editing; E.V.R.-F. Conceptualization, writing-review-editing; J.N. Conceptualization, formal analysis, writing—original draft preparation. All authors have read and agreed to the published version of the manuscript.

Funding: The authors acknowledge financial support by MINECO (Spain) through the projects MAT2017-86992-R and CTQ2017-88171-P, "Ministerio de Ciencia e innovación" (PID2020-116998RB-I00), Ministerio de Educación y Formación Profesional (PRX21/00407), and Conselleria de Innovacion, Universidades, Ciencia y Sociedad Digital (CIPROM/2021/022, MFA/2022/048).

Data Availability Statement: Not applicable.

Conflicts of Interest: The authors declare no conflict of interest.

Appendix A

Epichlorohydrin. ^{1}H-NMR (300 MHz, $CDCl_3$) δ = 3.67-3.54 (m, 1H), 3.50-3.44 (m, 1H), 3.26-3.15 (m, 1H), 2.85 (dd, *J* = 4.8, 3.9 Hz, 1H), 2.65 (dd, *J* =4.8, 2.5 Hz, 1H).

4-(Chloromethyl)-1,3-dioxolan-2-one. ^{1}H-NMR (300 MHz, $CDCl_3$) δ = 5.00-4.94 (m, 1H), 4.57 (dd, *J* = 8.9, 8.3 Hz, 1H), 4.37 (dd, *J* = 8.9, 5.7 Hz, 1H), 3.82-3.77 (m, 1H), 3.72-3.65 (m, 1H).

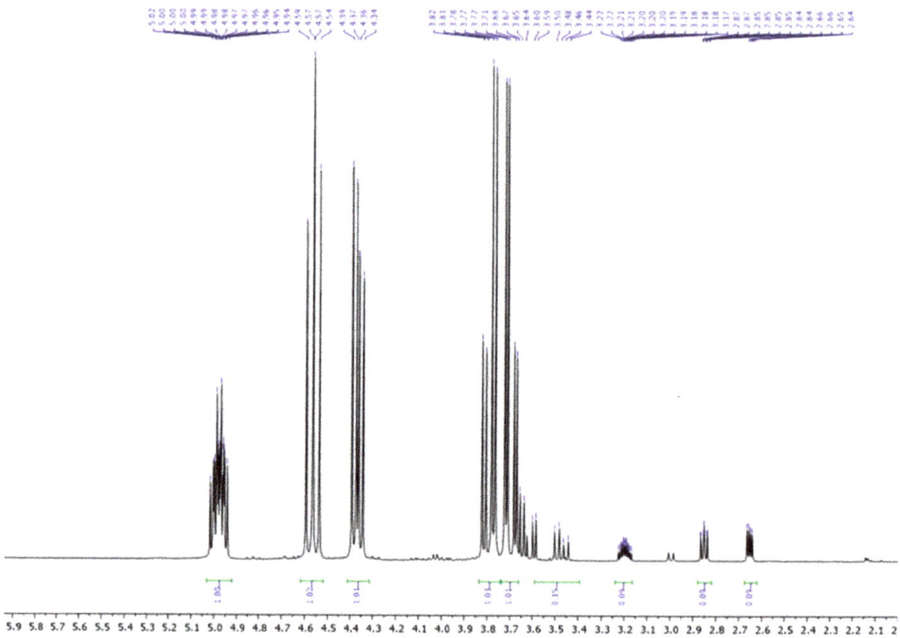

Figure A1. ^{1}H-NMR spectra of the reaction crude.

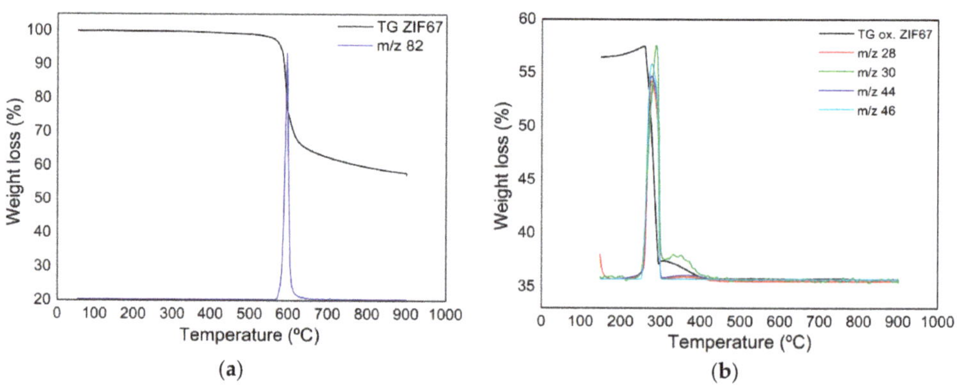

Figure A2. TGA-MS of sample ZIF-67. (**a**) In Ar, where *m/z* 82 is related with the 2-methylimidazole; (**b**) in 4:1 Ar:O_2, where *m/z* 28 is related to CO, *m/z* 30 is NO, *m/z* 44 is CO_2, and *m/z* is 46 NO_2.

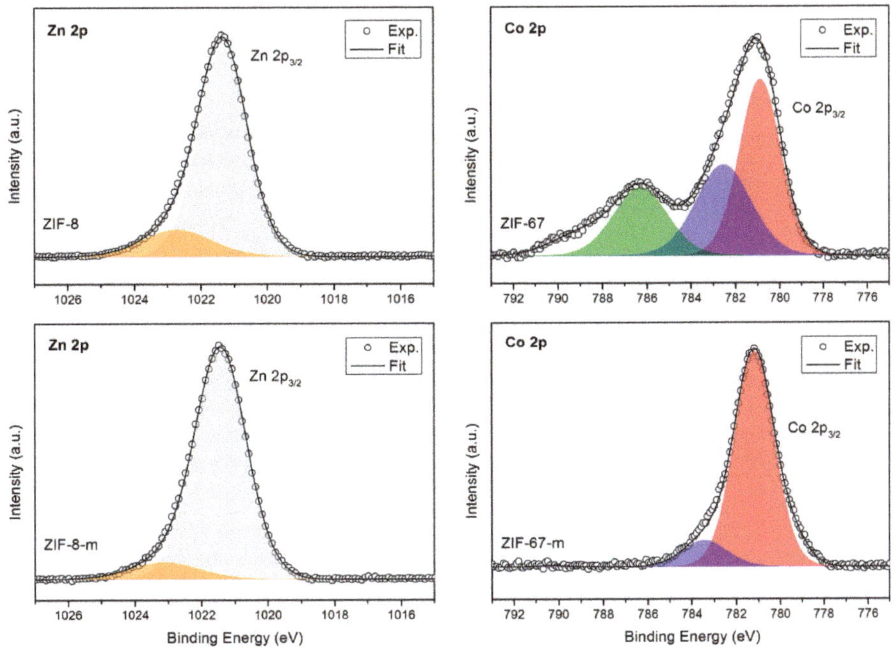

Figure A3. XPS of the four catalysts.

References

1. Pescarmona, P.P. Cyclic carbonates synthesised from CO_2: Applications, challenges and recent research trends. *Curr. Opin. Green Sustain. Chem.* **2021**, *29*, 100457. [CrossRef]
2. Kamphuis, A.J.; Picchioni, F.; Pescarmona, P.P. CO_2-fixation into cyclic and polymeric carbonates: Principles and applications. *Green Chem.* **2019**, *21*, 406–448. [CrossRef]
3. Maltby, K.A.; Hutchby, M.; Plucinski, P.; Davidson, M.G.; Hintermair, U. Selective Catalytic Synthesis of 1,2- and 8,9-Cyclic Limonene Carbonates as Versatile Building Blocks for Novel Hydroxyurethanes. *Chem.-A Eur. J.* **2020**, *26*, 7405–7415. [CrossRef] [PubMed]
4. Liu, J.; Li, Y.; Liu, H.; He, D. Transformation of CO_2 and glycerol to glycerol carbonate over CeO_2–ZrO_2 solid solution—Effect of Zr doping. *Biomass Bioenergy* **2018**, *118*, 74–83. [CrossRef]
5. Razali, N.; McGregor, J. Improving product yield in the direct carboxylation of glycerol with CO_2 through the tailored selection of dehydrating agents. *Catalysts.* **2021**, *11*, 138. [CrossRef]
6. Rozulan, N.; Halim, S.A.; Razali, N.; Lam, S.S. A review on direct carboxylation of glycerol waste to glycerol carbonate and its applications. *Biomass Convers. Biorefinery* **2022**, *12*, 4665–4682. [CrossRef]
7. Duan, C.; Yu, Y.; Hu, H. Recent progress on synthesis of ZIF-67-based materials and their application to heterogeneous catalysis. *Green Energy Environ.* **2022**, *7*, 3–15. [CrossRef]
8. Xiang, W.; Sun, Z.; Wu, Y.; He, L.; Liu, C. Enhanced cycloaddition of CO_2 to epichlorohydrin over zeolitic imidazolate frameworks with mixed linkers under solventless and co-catalyst-free condition. *Catal. Today* **2020**, *339*, 337–343. [CrossRef]
9. Zhou, K.; Mousavi, B.; Luo, Z.; Phatanasri, S.; Chaemchuen, S.; Verpoort, F. Characterization and properties of Zn/Co zeolitic imidazolate frameworks vs. ZIF-8 and ZIF-67. *J. Mater. Chem. A.* **2017**, *5*, 952–957. [CrossRef]
10. Kuruppathparambil, R.R.; Jose, T.; Babu, R.; Hwang, G.Y.; Kathalikkattil, A.C.; Kim, D.W.; Park, D.W. A room temperature synthesizable and environmental friendly heterogeneous ZIF-67 catalyst for the solvent less and co-catalyst free synthesis of cyclic carbonates. *Appl. Catal. B Environ.* **2016**, *182*, 562–569. [CrossRef]
11. Caló, V.; Nacci, A.; Monopoli, A.; Fanizzi, A. Cyclic carbonate formation from carbon dioxide and oxiranes in tetrabutylammonium halides as solvents and catalysts. *Org. Lett.* **2002**, *4*, 2561–2563. [CrossRef]
12. Mazo, P.; Rios, L. Carbonation of epoxidized soybean oil improved by the addition of water. *J. Am. Oil Chem. Soc.* **2013**, *90*, 725–730. [CrossRef]
13. Anderson, C.E.; Vagin, S.I.; Xia, W.; Jin, H.; Rieger, B. Cobaltoporphyrin-catalyzed CO_2/epoxide copolymerization: Selectivity control by molecular design. *Macromolecules* **2012**, *45*, 6840–6849. [CrossRef]

14. Sugimoto, H.; Kuroda, K. The cobalt porphyrin–Lewis base system: A highly selective catalyst for alternating copolymerization of CO_2 and epoxide under mild conditions. *Macromolecules* **2008**, *41*, 312–317. [CrossRef]
15. Udayakumar, S.; Lee, M.K.; Shim, H.L.; Park, S.W.; Park, D.W. Imidazolium derivatives functionalized MCM-41 for catalytic conversion of carbon dioxide to cyclic carbonate. *Catal. Commun.* **2009**, *10*, 659–664. [CrossRef]
16. Barbarini, A.; Maggi, R.; Mazzacani, A.; Mori, G.; Sartori, G.; Sartorio, R. Cycloaddition of CO_2 to epoxides over both homogeneous and silica-supported guanidine catalysts. *Tetrahedron Lett.* **2003**, *44*, 2931–2934. [CrossRef]
17. Aprile, C.; Giacalone, F.; Agrigento, P.; Liotta, L.F.; Martens, J.A.; Pescarmona, P.P.; Gruttadauria, M. Multilayered supported ionic liquids as catalysts for chemical fixation of carbon dioxide: A high-throughput study in supercritical conditions. *ChemSusChem* **2011**, *4*, 1830–1837. [CrossRef] [PubMed]
18. Cheng, W.; Chen, X.; Sun, J.; Wang, J.; Zhang, S. SBA-15 supported triazolium-based ionic liquids as highly efficient and recyclable catalysts for fixation of CO_2 with epoxides. *Catal. Today* **2013**, *200*, 117–124. [CrossRef]
19. Taherimehr, M.; Van de Voorde, B.; Wee, L.H.; Martens, J.A.; De Vos, D.E.; Pescarmona, P.P. Strategies for Enhancing the Catalytic Performance of Metal–Organic Frameworks in the Fixation of CO_2 into Cyclic Carbonates. *ChemSusChem* **2017**, *10*, 1283–1291. [CrossRef]
20. Zalomaeva, O.V.; Maksimchuk, N.V.; Chibiryaev, A.M.; Kovalenko, K.A.; Fedin, V.P.; Balzhinimaev, B.S. Synthesis of cyclic carbonates from epoxides or olefins and CO_2 catalyzed by metal-organic frameworks and quaternary ammonium salts. *J. Energy Chem.* **2013**, *22*, 130–135. [CrossRef]
21. Ramos-Fernández, E.V.; Serrano-Ruiz, J.C.; Sepúlveda-Escribano, A.; Narciso, J.; Ferrando-Soria, J.; Pardo, E. CHAPTER 9 Metal Organic Frameworks: From Material Chemistry to Catalytic Applications. In *Heterogeneous Catalysis for Energy Applications*; Reina, T.R., Odriozola, J.A., Eds.; The Royal Society of Chemistry: London, UK, 2020; pp. 235–303. ISBN 978-1-78801-718-3.
22. Delgado-Marín, J.J.; Izan, D.P.; Molina-Sabio, M.; Ramos-Fernandez, E.V.; Narciso, J. New Generation of MOF-Monoliths Based on Metal Foams. *Molecules* **2022**, *27*, 1968. [CrossRef] [PubMed]
23. Ronda-Lloret, M.; Pellicer-Carreño, I.; Grau-Atienza, A.; Boada, R.; Diaz-Moreno, S.; Narciso-Romero, J.; Serrano-Ruiz, J.C.; Sepúlveda-Escribano, A.; Ramos-Fernandez, E.V. Mixed-Valence Ce/Zr Metal-Organic Frameworks: Controlling the Oxidation State of Cerium in One-Pot Synthesis Approach. *Adv. Funct. Mater.* **2021**, *31*, 2102582. [CrossRef]
24. Ramos-Fernandez, E.V.; Redondo-Murcia, A.; Grau-Atienza, A.; Sepúlveda-Escribano, A.; Narciso, J. Clean production of Zeolitic Imidazolate Framework 8 using Zamak residues as metal precursor and substrate. *J. Clean. Prod.* **2020**, *260*, 121081. [CrossRef]
25. Beyzavi, M.H.; Stephenson, C.J.; Liu, Y.; Karagiaridi, O.; Hupp, J.T.; Farha, O.K. Metal-organic framework-based catalysts: Chemical fixation of CO_2 with epoxides leading to cyclic organic carbonates. *Front. Energy Res.* **2015**, *2*, 63. [CrossRef]
26. Ouellette, W.; Hudson, B.S.; Zubieta, J. Hydrothermal and Structural Chemistry of the Zinc(II)- and Cadmium(II)-1,2,4-Triazolate Systems. *Inorg. Chem.* **2007**, *46*, 4887–4904. [CrossRef] [PubMed]
27. Oullette, W.; Prosvirin, A.V.; Valeich, J.; Dunbar, K.R.; Zubieta, J. Hydrothermal Synthesis, Structural Chemistry, and Magnetic Properties of Materials of the M^{II}/Triazolate/Anion Family, Where M^{II} = Mn, Fe, and Ni. *Inorg. Chem.* **2007**, *46*, 9067–9082. [CrossRef] [PubMed]
28. Bondarenko, G.N.; Dvurechenskaya, E.G.; Ganina, O.G.; Alonso, F.; Beletskaya, I.P. Solvent-free Synthesis of Cyclic Carbonates from CO_2 and Epoxides Catalyzed by Reusable Alumina-supported Zinc Dichloride. *Appl. Catal. B: Environ.* **2019**, *254*, 380–390. [CrossRef]
29. Kuruppathparambil, R.R.; Babu, R.; Jeong, H.M.; Hwang, G.Y.; Jeong, G.S.; Kim, M.I.; Kim, D.W.; Park, D.W. A Solid SolutionZeolitic Imidazolate Framework as a Room Temperature Efficient Catalyst for the Chemical Fixation of CO_2. *Green Chem.* **2016**, *18*, 6349–6356. [CrossRef]
30. Mousavi, B.; Chaemchuen, S.; Moosavi, B.; Luo, Z.; Gholampour, N.; Verpoort, F. Zeolitic Imidazole Framework-67 as an Efficient Heterogeneous Catalyst for the Conversion of CO_2 to Cyclic Carbonates. *New J. Chem.* **2016**, *40*, 5170–5176. [CrossRef]
31. Chizallet, C.; Lazare, S.; Bazer-Bachi, D.; Bonnier, F.; Lecocq, V.; Soyer, E.; Quoineaud, A.A.; Bats, N. Catalysis of Transesterification by a Nonfunctionalized Metal−Organic Framework: Acido-Basicity at the External Surface of ZIF-8 Probed by FTIR and ab Initio Calculations. *J. Am. Chem. Soc.* **2010**, *132*, 12365–12377. [CrossRef]
32. Blackman, A. Reactions of Coordinated ligands. *Adv. Heterocycl. Chem.* **1993**, *58*, 123–170.
33. Cox, J.R.; Woodcock, S.; Hillier, I.H.; Vincent, M.A. Tautomerism of 1,2,3- and 1,2,4-Triazole in the Gas Phase and In Aqueous Solution: A Combined ab Initio Quantum Mechanics and Free Energy Perturbation Study. *J. Phys. Chem.* **1990**, *94*, 5499–5501. [CrossRef]
34. Lunazi, L.; Parisi, F.; Macciantelli, D. Conformational studies by dynamic nuclear magnetic resonance spectroscopy. Part 27. Kinetics and mechanism of annular tautomerism in isomeric triazoles. *J. Chem. Soc. Perkin Trans. 2* **1984**, *6*, 1025–1028. [CrossRef]
35. Garrat, P.J. *Comprehensive Heterocyclic Chemistry II*; Katrizky, A.R., Rees, C.W., Scriven, E.F.V., Storr, R.C., Eds.; Pergamon: Exeter, UK, 1996; Volume 4, p. 135.
36. Narciso, J.; Ramos-Fernandez, E.V.; Delgado-Marín, J.J.; Affolter, C.W.; Olsbye, U.; Redekop, E.A. New route for the synthesis of Co-MOF from metal substrates. *Microporous Mesoporous Mater.* **2021**, *324*, 111310. [CrossRef]
37. Villalgordo-Hernández, D.; Grau-Atienza, A.; García-Marín, A.A.; Ramos-Fernández, E.V.; Narciso, J. Manufacture of Carbon Materials with High Nitrogen Content. *Materials* **2022**, *15*, 2415. [CrossRef]

molecules

MDPI

Article

Biological Effects of EGCG@MOF $Zn(BTC)_4$ System Improves Wound Healing in Diabetes

ong Li [1,*,†], Jing Yan [2,3,†], Qiangqiang Zhu [2], Xinxiang Liu [2], Senlin Li [1], Shenhou Wang [1], Xuanjun Wang [1,2,*] and Jun Sheng [2,*]

1 College of Science, Yunnan Agricultural University, Kunming 650201, China
2 Key Laborotory of Puer Tea Sciencs, Ministry of Education(YNAU), Yunnan Agricultural University, Kunming 650201, China
3 Agro-Products Processing Research Institute, Yunnan Academy of Agricultural Sciences, Kunming 650201, China
* Correspondence: sirleee@live.cn (S.L.); jwang@ynau.edu.cn (X.W.); shengj@ynau.edu.cn (J.S.)
† These authors have contributed equally to this work.

Abstract: Tea contains high levels of the compound epigallocatechin gallate (EGCG). It is considered an important functional component in tea and has anti-cancer, antioxidant, and anti-inflammatory effects. The eight phenolic hydroxyl groups in EGCG's chemical structure are the basis for EGCG's multiple biological effects. At the same time, it also leads to poor chemical stability, rendering EGCG prone to oxidation and isomerization reactions that change its original structure and biological activity. Learning how to maintain the activity of EGCG has become an important goal in understanding the biological activity of EGCG and the research and development of tea-related products. Metal–organic frameworks (MOFs) are porous materials with a three-dimensional network structure that are composed of inorganic metals or metal clusters together with organic complexes. MOFs exploit the porous nature of the material itself. When a drug is an appropriate size, it can be wrapped into the pores by physical or chemical methods; this allows the drug to be released slowly, and MOFs can also reduce drug toxicity. In this study, we used MOF $Zn(BTC)_4$ materials to load EGCG and investigated the sustained release effect of EGCG@MOF $Zn(BTC)_4$ and the biological effects on wound healing in a diabetic mouse model.

Keywords: EGCG; MOF; wound healing; diabetes

1. Introduction

Tea is one of the most popular drinks in the world—far exceeding beer, coffee, and carbonated drinks— and is well known in the forms of green, black, or oolong tea. Among these forms, green tea has had the most obvious impact on human health [1–4]. The production of green tea is to cut, twist, and dry freshly picked young tea tree leaves to produce a dry and stable product. The enzymes responsible for decomposing the pigments in the young leaves are destroyed in the process of curing, so the leaves remain green during the subsequent rolling and drying process. This production method preserves the natural polyphenol content in tea, which are the compounds responsible for its biological activity. Most green tea polyphenols (GTP) are flavonols, usually called catechins. There are four main types of catechins in green tea: epicatechin (EC), epicatechin gallate (ECG), epigallocatechin(EGC), and epigallocatechingallate (EGCG). The most abundant is EGCG [5–7].

The ester compound synthesized by the combination of gallocatechol and gallic acid is EGCG, which has a molecular weight of 453 kDa. EGCG is a four-ring structure containing eight phenolic hydroxyl groups. These structural characteristics imbue strong antioxidant properties. Studies have shown that EGCG can directly inhibit the expression of inflammatory factors such as tumor necrosis factor (TNF)-α, interleukin (IL)-6, and IL-1β, thereby achieving anti-inflammatory effects. EGCG undergoes auto-oxidation to generate hydrogen peroxide that has long-term antioxidant activity [8,9]. EGCG can combat the harmful effects of many potentially pathogenic bacteria [10–12]. For example, EGCG is very effective in inhibiting the growth of Staphylococcus aureus and methicillin-resistant Staphylococcus aureus (MRSA) [13–15]. Among catechins, EGCG reacts effectively with most ROS (reactive oxygen species). However, due to its unique chemical structure, EGCG itself is also very unstable and is easily affected by factors such as temperature, metal ions, enzymes, and pH. It is prone to oxidation and isomerization reactions, thereby changing its original structure and biological activity.

At present, protein nanoparticles, polysaccharide nanoparticles, liposomes, etc., are used as delivery bodies to encapsulate drugs, which can prevent the isomerization and oxidation reaction of EGCG, improve its slow release and stability, and realize targeted delivery, thus expanding the application range of EGCG. However, EGCG has poor lipid solubility, resulting in poor permeability in cells and low absorption rate, resulting in a low oral utilization rate of liposomes in clinical treatment. In addition, liposomes are easily damaged by a weak acid environment and digestive enzymes in the process of digestion, resulting in the leakage of the load. Therefore, liposomes as delivery carriers of EGCG still need further research. The stability of EGCG-CS nanoparticles in the intestine was affected by the porosity of polysaccharides, low embedding rate, and core material [16].

In recent years, porous materials have attracted much attention in the fields of materials, chemistry, physics, and others because of their simple preparation process, large specific surface area, easy-to-optimize structure, and orderly porous structure. Metal–organic frameworks (MOFs) are a new type of porous material that combines inorganic and organic components through physical or chemical action. MOF materials not only have a crystal structure similar to regular pores of zeolite molecular sieve but also have a higher specific surface area than traditional porous materials such as polymer, an inorganic polymer, and carbon-based materials. In addition, it contains organic components, which makes its structure tailorable, designable, adjustable pore size, and functional pore surface. MOF materials have a rich structure, and the hole size and type can be designed according to project needs. MOFs could be used to develop slow-release drugs and even address environmental pollution and energy issues [17,18].

MOF materials are prepared by mixing metal ions with organic complexes in a solvent, heating the mixture to generate and grow crystal nuclei, and combining the metal ions with small molecular substances in the solvent to form a competitive relationship with the organic complexes. After washing and drying, the small molecules of the solvent can be separated from the metal sites, thereby forming unsaturated metal sites, which enables adsorption or catalysis. The MOF material itself has a porous structure. When the size of the pores is larger than a drug, that drug can be loaded into the pore by physical or chemical means. Once in a specific environment, the drug can be released slowly over time, which can reduce drug toxicity and side effects.

MOF $Zn(BTC)_4$ is a new MOF material synthesized by our research group. It was chosen as a model MOF because its synthesis was simple and scalable, and the toxicity of the ligand H3BTC was reported to be low [19].

The most serious complications of diabetes occur after long-term treatment. In clinical research, the development of chronic inflammation is an important sign of spontaneous diabetes. The level of inflammation is correlated with diabetes complications [20]. Slow wound healing is one of the common characteristics of subjects with diabetes. Wounds in these patients are difficult to heal due to chronic inflammation and high blood sugar [21]. The presence of chronic inflammation in diabetic wounds is manifested by an increase in pro-inflammatory factors [22]. Compared with normal wounds, diabetic wounds exhibit obvious high expression of inflammatory factors, including IL-1β, TNF-α, and others [23].

EGCG itself has a strong anti-inflammatory effect and can promote wound healing in diabetic mice [24]. However, because of the unsaturated bond in the structure of EGCG, its chemical stability is poor, and the instability of EGCG also leads to a decrease in its bioavailability, which limits its application in the food and pharmaceutical industries. Therefore, in this study, the streptozotocin (STZ)-induced diabetic mouse model was used as the main research object to explore the effects of EGCG and EGCG@MOF $Zn(BTC)_4$ on wound healing in diabetic mice. We explored whether EGCG@MOF $Zn(BTC)_4$ can normally release EGCG, reduce its degradation rate, and prolong its biological effects [25].

2. Result and Discussion

2.1. General Information

The following materials were obtained from the indicated vendors: tribenzoic acid (Shanghai Darui Fine Chemicals Co., Ltd. Shanghai, China), zinc nitrate (Aladdin), absolute ethanol (Tianjin Fengchuan Chemical Reagent Technology Co., Ltd. Tianjin, China), N,N-dimethylamide (Sinopharm Chemical Reagent Co., Ltd. Company, Shanghai, China), methanol (Tianjin Fengchuan Chemical Reagent Technology Co., Ltd.), EGCG (purity above 98%) (Chengdu Purifa Technology Development Co., Ltd. Chengdu, China), RAW264.7 cells (Kunming Institute of Animal Cell Bank), lipopolysaccharide (Sigma), TNF-α kit (Beijing Sizhengbai Biological Co., Ltd. Beijing, China), STZ (Sigma, St Louis, MO, USA), ICR mice (Changzhou Cavens, Changzhou, China), sodium carboxymethyl cellulose (Aladdin), Ma Song (Masson) Staining kit (Beijing Zhongshan Jinqiao Biotechnology Co., Ltd. Beijing, China), IL-1β kit (Beijing Sizhengbai Biotechnology Co., Ltd. Beijing, China), and IL-6 kit (Beijing Sizhengbai Biotechnology Co., Ltd. Beijing, China). The data for crystals were collected on a Bruker APEX-II CCD diffractometer (MoKa, l 1/4 0.71073 Å) at 298(2) K. Infrared spectra (KBr pellets) were taken on a Bruke Tensor 27 FTIR spectrometer in the range of 4000–400 cm^{-1}. Morphology was recorded on a FlexSEM 1000 scanning electron microscope. Microscopic characterization was finished by Tianqing microscope (optical magnification: 50×–1000×, accuracy: 0.001 mm).

2.2. Synthesis of MOF $Zn(BTC)_4$

Dissolve $Zn(NO_3)_2{\bullet}6H_2O$ in 60 mL of ultrapure water to yield a concentration of 6.7 mmol, stir it evenly, add methanol and trimesic acid on this basis, and finally heat it in an oven at 100 °C for 24 h. After the heating process is over, the oven is kept closed until the temperature is naturally cooled down to room temperature. The cooled sample was filtered; cleaned; and dried under vacuum at 60 °C to obtain white, rectangular, powdery crystals. Finally, the sample was collected, sealed, and stored for later use.

2.3. Preparation of EGCG@MOF

Weigh 20 mg of MOF $Zn(BTC)_4$ into a 2 mL Eppendorf (EP) tube, add 1 mL of 140 mg/mL EGCG solution, mix well, place in a shaker at 4 °C, and rotate slowly for 12 days. Afterward, take out the EP tube and centrifuge at 4 °C, 4500 r/min for 10 min, discard the supernatant as much as possible, add 2 mL of absolute ethanol to resuspend EGCG@MOF, repeat the centrifugation twice, discard both supernatants, and wash it as much as possible to remove the remaining EGCG on the surface of the MOF. Add 1 mL of anhydrous ethanol–water solution to the EP tube, place it in a freezer at −80 °C until frozen, then place it in a freeze dryer to produce dry EGCG@MOF powder.

2.4. Detection of Drug Loading of EGCG@MOF

Accurately weigh the mass of EGCG@MOF in the EP tube, add 1 mL of acidified phosphate-buffered saline (PBS) solution to resuspend it, and place it in a 37 °C water bath for ultrasound for 1 h to completely release the EGCG in the MOF. Filter it with a 0.45 μm pore size filter, use high-performance liquid chromatography (HPLC) to detect the concentration of EGCG, and calculate the drug loading amount of EGCG for the MOF.

2.5. HPLC Detection of EGCG Concentration

A sample filtered through a 0.45 μm pore was used to determine the EGCG content in the sample using HPLC. The platform was equipped with G1311B Quat pump (1 mL per minute), G1329B autosampler, G1316A TCC column oven (40 °C), and G1314F ultraviolet detector (280 nm) (Agilent 1260 series HPLC system, Agilent, Santa Clara, CA, USA). The sample volume was 10 μL, and the HPLC experiment was controlled with 1260LC Agilent ChemStation software (Agilent Technologies). A C18 ODS column (ZORBAX SB-C18 4.6 mm 250 mm, 5 μm, Agilent) was used to complete the separation. The mobile phases used in the HPLC analysis were solvents A (100% acetonitrile) and B (0.03% trifluoroacetic acid), which were filtered through a 0.45 μm filter. Within 25 min, mobile phase A increased from 10% to 60%.

2.6. Cell Processing and Grouping

RAW264.7 cells were digested, counted with a cell counter, and inoculated with 10 mL of culture medium at the number of five million cells per plate. Then the plates were put in a 37 °C incubator for 24 h. Afterward, the culture medium was discarded, the cells were washed once with PBS, and serum-free high-glycemic culture medium was added for starvation conditions and cultured overnight.

After overnight starvation, the cells were given fresh serum-free medium, and all EGCG@MOF $Zn(BTC)_4$ groups were added according to the release concentration of EGCG to a final concentration of 10 μg/mL and placed in a 37 °C incubator for 30 min. The dosage of the MOF $Zn(BTC)_4$ group was added according to EGCG@MOF $Zn(BTC)_4$ group. After 30 min, lipopolysaccharide (LPS) solution with a concentration of 200 EU/mL was added to the culture medium, and the plates were placed in a carbon dioxide incubator at 37 °C for another 3 h and 30 min before collecting the cell supernatant and extracting the protein. There were five groups of cells used in the experiments: Control, LPS, MOF $Zn(BTC)_4$, EGCG@MOF $Zn(BTC)_4$, and LPS+EGCG@MOF $Zn(BTC)_4$.

2.7. Establishment of Diabetic Mouse Model

All ICR mice fasted for 15 h with free access to water. The mice were then weighed using an electronic scale, the average weight was calculated, and the cages were randomly divided according to the rules of consistent weight average. According to our preliminary studies and reference to other experimental methods, a 1% STZ solution was prepared for intraperitoneal injection of ICR mice at a dose of 100 mg/kg. The injected mice were placed back into the cage, allowed to eat and drink freely, and raised normally. Their blood glucose levels were tested once a week for 14 days. Reaching a blood glucose level >20 mmol/L and then maintaining it for more than 2 weeks was considered a successful model for diabetic mice.

2.8. Establishment of a Wound Healing Model

The successful ICR diabetic mouse model and normal ICR mice were used as the research objects for the skin wound model. After using an electric razor to completely shave the dorsal fur, a depilatory cream was applied to completely remove the remaining hair. After rinsing off the depilatory cream with warm water, the skin was dried with absorbent paper, and the mice were placed back in the cage under a warm lamp. Because the depilatory cream can irritate mouse skin of the mice, the wounds were made the next day.

The mice were anesthetized by inhaling ether from their mouths. The dorsal area was washed with 75% ethanol, and an 8 mm biopsy device was used to remove a piece of skin on the left and right sides of the mouse's back. A steel ruler was used as an indicator to mark the areas, treat the wounds with the compounds, and determine how photos were taken for the experimental records.

2.9. Wound Healing Experiment Grouping and Treatment

There were five experimental groups: normal, diabetes, MOF $Zn(BTC)_4$, EGCG@MOF $Zn(BTC)_4$, and EGCG. The normal group consisted of control ICR mice, and the other four groups were randomly divided diabetic model mice (7 mice/group). All mice were reared under the same environmental conditions with ad libitum access to food and water. All wound model mouse dressings are prepared with 1% carboxymethyl cellulose (CMC) solution. Diabetic and normal mice were treated with 1% CMC solution as a placebo. Mice in the EGCG@MOF $Zn(BTC)_4$ group were treated with EGCG@MOF $Zn(BTC)_4$ dressings with an EGCG concentration of 20 μg/mL. The amount of MOF $Zn(BTC)_4$ groups added is equivalent to EGCG@MOF $Zn(BTC)_4$ group addition amount. The mice in the EGCG group were treated with 20 μg/mL EGCG.

The drugs in all treatment groups were given once every 4 days. The experimental period was 15 days. All treatments were given at 10 a.m., and the wounds were photographed and measured every day. The images were analyzed ImageJ software (National Institutes of Health) to measure wounds, and these data were used to calculate the healing rate and determine if there were significant differences between the different treatment groups.

The wound healing rate was calculated as: 1–wound area on day n(1,2,3 ...)/initial wound area.

2.10. Wound Sampling Treatment

Depending on the results of the wound healing experiment in mice, different time points were selected for tissue extraction with an 8 mm biopsy sampler. After tissue collection, the samples were washed in cold PBS to remove blood, dried with filter paper to remove surface water, and then half were placed in formalin for at least 24 h. The other half were placed in liquid nitrogen for quick freezing, and then transferred to an −80 °C freezer for storage until subsequent experiments. Mouse skin tissue preserved in liquid nitrogen is used for enzyme-linked immunosorbent assay to detect related inflammatory factors.

3. Results and discussion

3.1. Structural Characterization of MOF $Zn(BTC)_4$

The structure of MOF $Zn(BTC)_4$ was characterized by crystal diffraction. The crystallographic data of MOF $Zn(BTC)_4$ are listed in Table 1, and the selected bond lengths and angles are listed in Table 2. Figure 1 respectively shows the SBU image and 3D frame structure of this crystal.

Table 1. Crystal data of MOF $Zn(BTC)_4$.

Empirical Formula	$C_{36}H_{24}O_{24}Zn$	Z	4
Formula weight	809.67	μ/nm^{-1}	1.403
Size/mm	0.17 × 0.15 × 0.10	$Dc/(mg.cm^{-3})$	1.449
θ rang for data collection/(°)	2.894 to 26.085	F(000)	808
Crystal system	Monoclinic	Reflections collected	9040
Space group	P2(1)/n	V/nm^3	1795.7(2)
a/Å	9.5077(5)	Goodness of fit on F^2	1.010
b/Å	16.3950(16)	$R1,wR_2(I > 2\sigma(I)$	0.0456, 0.1080
c/Å	11.6119(9)	$R1,wR_2$(all data)	0.0683, 0.1172
α/(°)	90.00	$\Delta\rho max(eÅ^{-3})$	0.797
β/(°)	97.2200(10)	$\Delta\rho min(eÅ^{-3})$	−0.544
γ/(°)	90.00		

Table 2. Selected bond lengths(Å) and angles(°) for MOF $Zn(BTC)_4$.

Zn1 O5	**1.939(3)**	**Zn1 O3**	**1.939(3)**	**Zn1 O2**	**1.967(3)**
Zn1 O1	1.972(3)				
O5 Zn1 O3	125.79(12)		O5 Zn1 O2	94.56(12)	
O3 Zn1 O2	113.35(14)		O5 Zn1 O1	109.99(12)	
O3 Zn1 O1	99.47(12)		O2 Zn1 O1	114.57(12)	

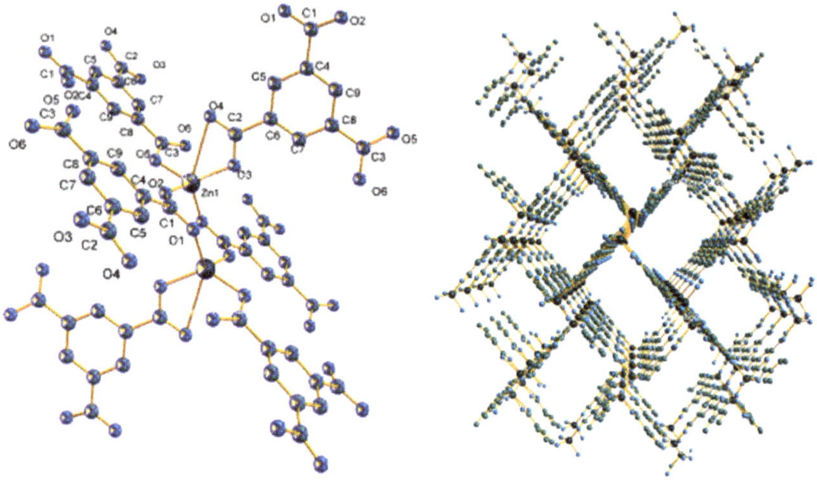

Figure 1. SBU image and 3D framework of MOF $Zn(BTC)_4$.

CCDC: 1539458 for compound 1.

3.2. Load Capacity of MOF Zn(BTC)$_4$ to Hold EGCG

Due to MOF's dual characteristics of high drug loading and controllable drug release, it has quickly become a research hotspot. Studies have shown that a small amount of mesoporous MOFs can carry a large drug load, and MOFs with larger pore diameters have more application value [26]. Multi-level MOFs with micropores, mesopores, and macropores can simultaneously carry out drug molecules of different sizes [27].

We first determined the peak of EGCG as a reference standard with HPLC, and the peak time of EGCG was 9.910 min (Figure 2A). After soaking MOF $Zn(BTC)_4$ in 50 mg/mL EGCG solution, the sample was rotated on a shaker at 4 °C for 8 days, then centrifuged twice at 4 °C with absolute ethanol to clean the surface of MOF from residual EGCG, and then ethanol was used to resuspend in an aqueous solution prior to freeze-drying to produce MOF material loaded with EGCG. The loaded EGCG@MOF was added to PBS and ultrasonicated at 37 °C for 1 h to completely release the EGCG in the MOF material. After filtration, the EGCG content was detected by HPLC. As shown in Figure 2B, the EGCG peak existed in the liquid phase spectrum of MOF $Zn(BTC)_4$ at 9.856 min, indicating that MOF $Zn(BTC)_4$ can load EGCG. Under these conditions, EGCG@MOF $Zn(BTC)_4$ loading was 56.247 mg/g.

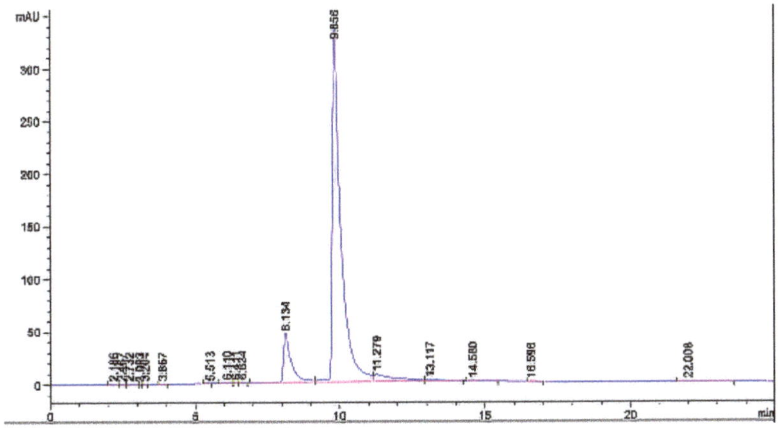

(A)

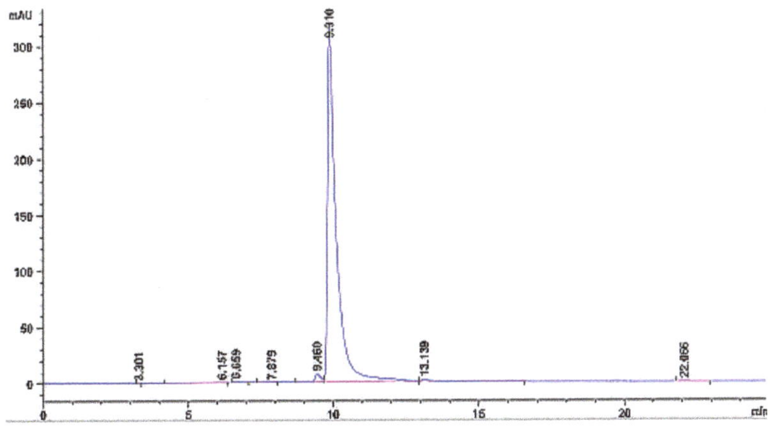

(B)

Figure 2. HPLC(High Performance Liquid Chromatography) detection of EGCG. (**A**) HPLC spectrum of EGCG standard; (**B**) MOF $Zn(BTC)_4$ loading of EGCG.

Next, we optimized EGCG loading into MOF $Zn(BTC)_4$ to EGCG. To increase drug loading, we performed a time course experiment. We weighed several pieces of 20 mg of MOF $Zn(BTC)_4$ into a 2 mL EP tube, added 1 mL of 50 mg/mL EGCG, mixed it upside down, placed it in a 4 °C shaker, and rotated it slowly. Every two days, one piece was removed, centrifuged, washed to remove surface EGCG, then frozen and dried to obtain MOF $Zn(BTC)_4$-EGCG. Next, 1 mL PBS was added before ultrasonication for 1 h. The resulting solution was filtered prior to HPLC detection of EGCG to calculate the amount of EGCG that can be loaded by MOF $Zn(BTC)_4$. The amount of EGCG loaded in EGCG@MOF $Zn(BTC)_4$ increased with time and reached the maximum on the 12th day (Figure 3). The highest loading amount of EGCG in EGCG@MOF $Zn(BTC)_4$ reached 94.134 mg/g on day 12 and then decreased, which means that EGCG@MOF $Zn(BTC)_4$ loaded with EGCG gradually accumulated with the increase in time, when the cumulative EGCG reached the maximum, the loaded EGCG will fall off and then eventually re-accumulate. Therefore, the amount of EGCG in MOF $Zn(BTC)_4$-EGCG will fluctuate within a certain range.

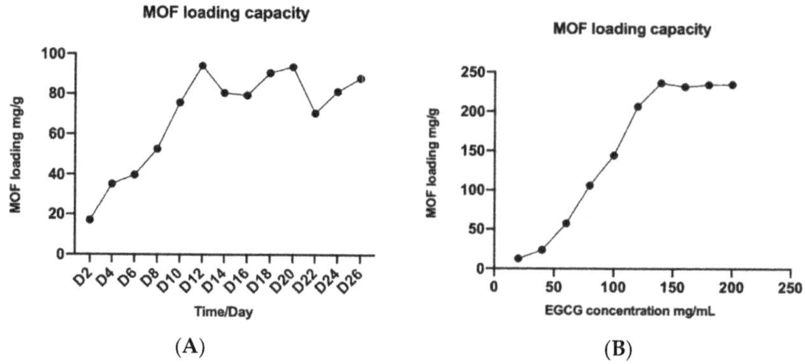

Figure 3. Optimizing MOF Zn(BTC)$_4$-EGCG loading. (**A**) The loading capacity of MOF Zn(BTC)$_4$ on EGCG over time; (**B**) The influence of different EGCG concentrations on EGCG@MOF Zn(BTC)$_4$ loading capacity.

To further increase the loading capacity of EGCG@MOF Zn(BTC)$_4$, we prepared different concentrations of EGCG (20, 30, 40, . . . , 200 mg/mL) for EGCG@MOF Zn(BTC)$_4$ loading. Briefly, we weighed 20 mg of EGCG@MOF Zn(BTC)$_4$ into a 2 mL EP tube, added 1 mL of EGCG of different concentrations, and placed the tubes on a shaker at 4 °C with slow rotation for 12 days. They were then centrifuged to wash away residual EGCG on the surface of EGCG@MOF Zn(BTC)$_4$, followed by freezing and freeze-drying to obtain EGCG@MOF Zn(BTC)$_4$. As described above, 1 mL PBS was added prior to 1 h ultrasonication. After filtering the solution, HPLC was performed to detect the concentration of EGCG and the amount of EGCG loaded in MOF Zn(BTC)$_4$. The amount of EGCG loaded in EGCG@MOF Zn(BTC)$_4$ varied with EGCG concentration (Figure 3B). When the EGCG concentration was 140 mg/mL, EGCG@MOF Zn(BTC)$_4$ reached a maximum of 236.227 mg/g.

3.3. Chemical Characterization of MOF Zn(BTC)$_4$ and EGCG@MOF Zn(BTC)$_4$

We observed MOF Zn(BTC)$_4$ and EGCG@MOF Zn(BTC)$_4$ powder grossly and microscopically and analyzed the morphological changes of the MOF Zn(BTC)$_4$ powder after loading EGCG.

In Figure 4A, the left and right sides show MOF Zn(BTC)$_4$ and EGCG@MOF Zn(BTC)$_4$ powder, respectively. MOF Zn(BTC)$_4$ is a white powdery solid, while MOF Zn(BTC)$_4$ is soaked in an EGCG solution and then freeze-dried. The color becomes slightly yellow after this process. MOF Zn(BTC)$_4$ itself is relatively finely broken and loose, and the powder form is an acicular solid when observed by the naked eye. When piled together, it appears relatively dense, while MOF Zn(BTC)$_4$ is soaked in EGCG solution. The color turns to pale yellow, the powder state has been destroyed, and the needle-like shape observed with the naked eye becomes more finely broken. When piled together, it appears fluffy; when packed in an EP tube, it is prone to hang on the walls of the tube. Figure 4B shows microscope images of the MOF Zn(BTC)$_4$ and EGCG@MOF Zn(BTC)$_4$ powders on the left and right, respectively. The MOF Zn(BTC)$_4$ powder is clearly visible and relatively dispersed with a fusiform structure. After filling the MOF Zn(BTC)$_4$ powder with EGCG and then freeze-drying, the original fusiform structure of MOF Zn(BTC)$_4$, becomes more finely fragmented and is easier to pile together. The solid materials are crushed and easily accumulate with obvious adhesion.

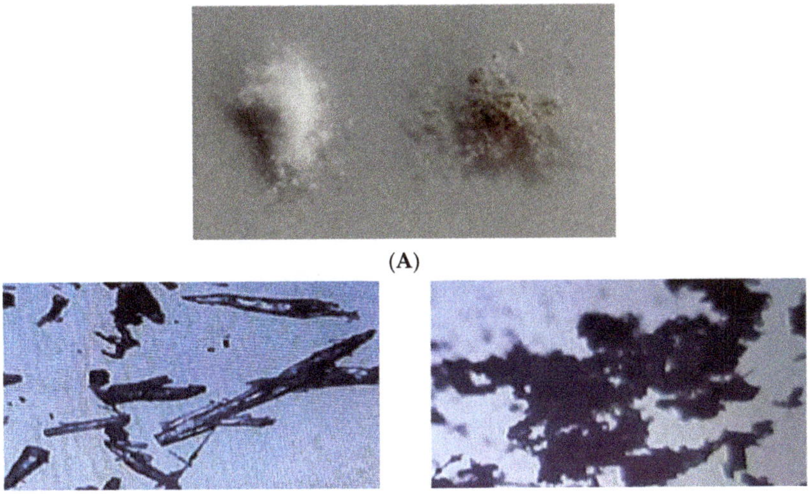

(A)

(B)

Figure 4. Characterization of MOF $Zn(BTC)_4$ and EGCG@MOF $Zn(BTC)_4$. (**A**) Gross images of MOF $Zn(BTC)_4$ and EGCG@MOF $Zn(BTC)_4$; (**B**) Microscope images of MOF $Zn(BTC)_4$ and EGCG@MOF $Zn(BTC)_4$ (40×).

The infrared (IR) spectrum analysis comparison of EGCG@MOF $Zn(BTC)_4$ and MOF $Zn(BTC)_4$ is shown in Figure 5A. The vibration peaks at 3387 cm^{-1}, 1619 cm^{-1}, 1557 cm^{-1}, 1443 cm^{-1}, 1372 cm^{-1}, 761 cm^{-1}, and 721 cm^{-1} did not change. However, compared with the absorption peak at 1110 cm^{-1} in the MOF $Zn(BTC)_4$ spectrum, new absorption peaks appeared at 1127 cm^{-1}, 1145cm^{-1}, and 1059 cm^{-1} in the EGCG@MOF $Zn(BTC)_4$ spectrum. This may be the stretching vibration generated by the C-O single bond in EGCG. These new absorption peaks likely appeared because introducing EGCG into MOF $Zn(BTC)_4$ changed the IR spectrum.

MOF $Zn(BTC)_4$ and EGCG@MOF $Zn(BTC)_4$ morphologies were assessed by scanning electron microscopy (SEM) at different magnifications (Figure 5B). MOF $Zn(BTC)_4$ crystals had a massive appearance, while the EGCG@MOF $Zn(BTC)_4$ crystal was more finely fragmented. After filling with EGCG, the structure, size, and crystal arrangement of MOF $Zn(BTC)_4$ were altered to certain degrees. As shown in the X-ray diffraction (XRD), results of MOF $Zn(BTC)_4$ and EGCG@MOF $Zn(BTC)_4$ were also compared (Figure 5C). When MOF $Zn(BTC)_4$ was filled with EGCG, the intensities of some characteristic peaks of MOF $Zn(BTC)_4$ changed significantly, but the peak shapes of MOF $Zn(BTC)_4$ were basically the same. After loading, the skeleton structure of MOF $Zn(BTC)_4$ was basically maintained.

PBS was used to prepare 0.5 mg/mL EGCG that was distributed into several 2 mL EP tubes, then 1 mL EGCG solution was added to each EP tube, and PBS was used to prepare 0.5 mg/mL EGCG@MOF $Zn(BTC)_4$ according to the EGCG concentration. Next, 1 mL of EGCG@MOF $Zn(BTC)_4$ solution was dispensed into several 2 mL EP tubes, the tubes were placed in a 37 °C drying oven, and 1 EP tube per group was removed every hour for 12 h. After that, one EP tube was taken out every 2 h for each group. Following filtration, the concentration of EGCG was detected by HPLC (Figure 6). The concentration of EGCG gradually decreased over time. At 24 h, EGCG has basically been completely degraded at a very fast rate. Conversely, EGCG@MOF $Zn(BTC)_4$ can gradually release EGCG and maintain the concentration within a certain range. The concentration of EGCG released by EGCG@MOF $Zn(BTC)_4$ also increased as a whole. It gradually decreased over time but to a much smaller degree than EGCG. At 120 h, the concentration of EGCG released by EGCG@MOF $Zn(BTC)_4$ was still maintained at 0.27 mg/mL.

At 10 h, the concentration of EGCG released by EGCG@MOF Zn(BTC)$_4$ was 0.38 mg/mL, and the concentration of EGCG alone was 0.27 mg/mL. At 20 h, the corresponding values were 0.49 and 0.13 mg/mL. At 26, 28, and 30 h, the contents of EGCG fluctuated because of the difference in the content of EGCG@MOF Zn(BTC)$_4$ in each EP tube. It can be clearly seen from the figure that the concentration of EGCG decreased linearly with time in both groups, but MOF Zn(BTC)$_4$ can stabilize the concentration of EGCG to a certain extent and slow its degradation. This may be due to the existence of pores in MOF Zn(BTC)$_4$ itself, which protects EGCG to a certain extent. Based on these results, we can conclude that EGCG@MOF Zn(BTC)$_4$ can slow down the degradation rate of EGCG, thereby prolonging its action time.

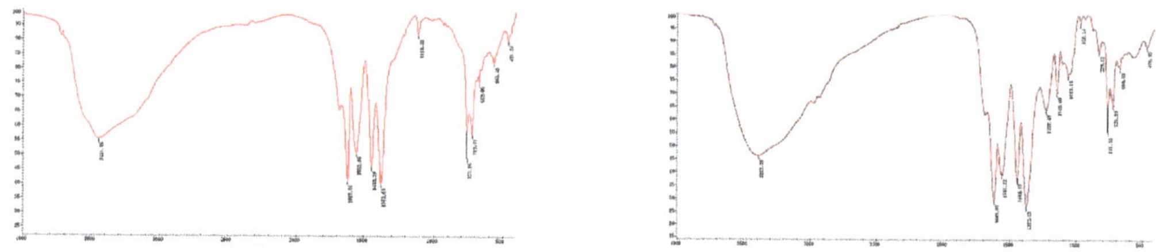

(**A**) MOF Zn(BTC)$_4$ EGCG@MOF Zn(BTC)$_4$

(**B**)

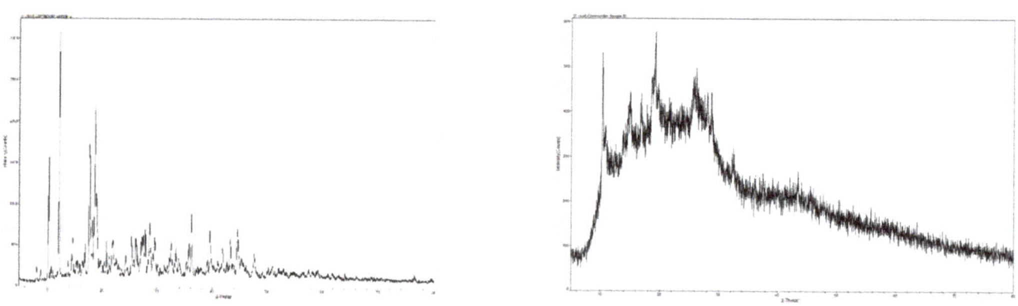

(**C**)

Figure 5. Comparison of MOF Zn(BTC)$_4$ and EGCG@MOF Zn(BTC)$_4$. (**A**) IR spectroscopy analysis of MOF Zn(BTC)$_4$ and EGCG@MOF Zn(BTC)$_4$; (**B**) Scanning electron microscopy (SEM) of MOF Zn(BTC)$_4$ and EGCG@MOF Zn(BTC)$_4$; (**C**) X-ray diffraction (XRD) analysis of MOF Zn(BTC)$_4$ and EGCG@MOF Zn(BTC)$_4$.

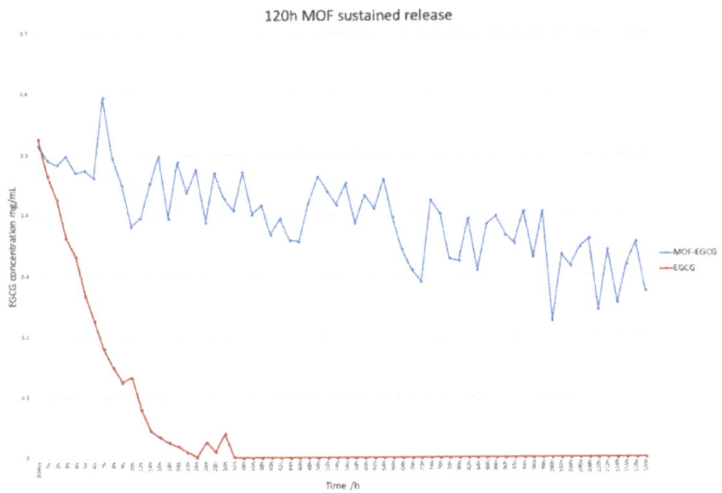

Figure 6. EGCG@MOF Zn(BTC)$_4$ slows EGCG degradation.

3.4. Research on Anti-Inflammatory Effects of EGCG@MOF Zn(BTC)$_4$ Targeting the Notch Signaling Pathway

The inflammatory response produced by macrophages can be attenuated by regulating Notch signaling, which further regulates the state of macrophages. Researchers have shown that blocking Notch1 protein production can reduce the activation level of macrophages [28]. We performed western blotting (WB) to detect the effect of EGCG on Notch protein secretion after 30 min and 3 h to assess whether EGCG can inhibit macrophage secretion of Notch-related proteins.

After LPS stimulation, the expression level of Notch1 protein expressed in macrophages did not increase significantly (Figure 7A). However, the amounts of Notch1 protein expressed in the EGCG@MOF Zn(BTC)$_4$+LPS and EGCG@MOF Zn(BTC)$_4$ groups were significantly lower than the control group after 3 h. The results indicate that EGCG@MOF Zn(BTC)$_4$ can inhibit the expression of Notch1 protein over time, but there was no obvious change after 30 min.

Notch2 protein expression in macrophages was significantly increased after LPS stimulation at 30 min and 3 h stimulation. After 30 min incubation with EGCG@MOF Zn(BTC)$_4$, Notch2 protein levels were not dramatically changed, but they were significantly reduced after 3 h stimulation. Macrophages treated with EGCG@MOF Zn(BTC)$_4$+LPS had higher Notch2 protein levels than those treated with EGCG@MOF Zn(BTC)4, indicating that EGCG@MOF Zn(BTC)$_4$ can dampen inflammation in macrophages over time.

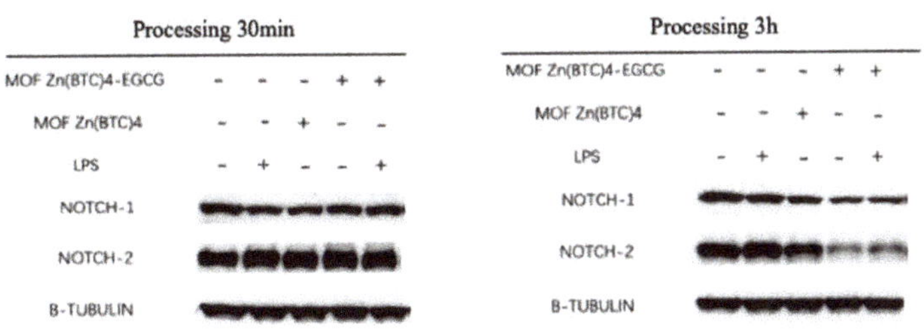

(A)

Figure 7. *Cont.*

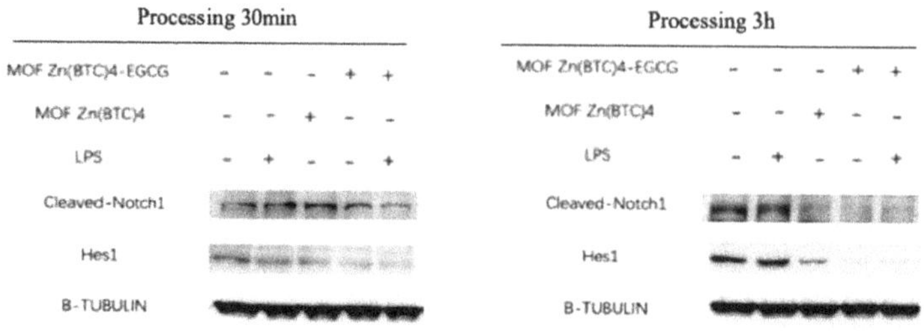

(B)

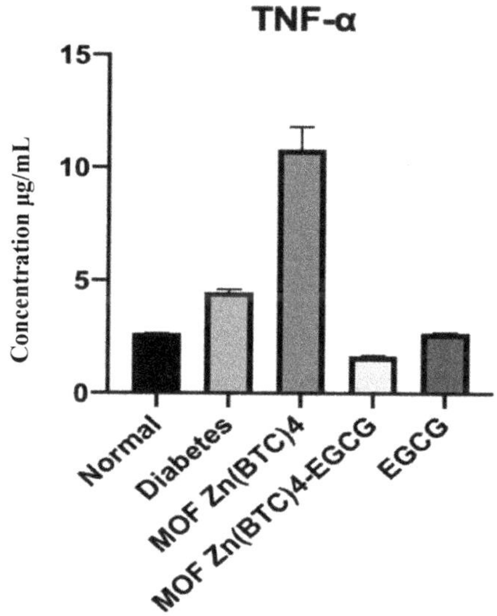

(C)

Figure 7. Effect of EGCG@MOF Zn(BTC)$_4$ on Notch protein expression and inflammatory factors in macrophages. (**A**) Effect of EGCG@MOF Zn(BTC)$_4$ on Notch protein expression at different time points; (**B**) Effect of EGCG@MOF Zn(BTC)$_4$ on Notch protein expression at different time points; (**C**) Effect of EGCG@MOF Zn(BTC)$_4$ on inflammatory factor levels.

To further illustrate the effect of EGCG@MOF Zn(BTC)$_4$ on Notch signaling, we once again used WB to detect the effect of EGCG@MOF Zn(BTC)$_4$ on Notch signaling pathway-related proteins at 30 min and 3 h. After 30 min of LPS treatment, Cleaved-Notch1 levels increased; it can be clearly seen that after treatment with EGCG@MOF Zn(BTC)$_4$, the expression levels of Cleaved-Notch1 and Hes1 proteins in macrophages were significantly reduced (Figure 7B). EGCG@MOF Zn(BTC)$_4$ had a significant inhibitory effect on Cleaved-Notch1 and Hes1 protein levels. When comparing the 30 min and 3 h timepoints, Hes1 protein levels increased with time. Moreover, after EGCG@MOF Zn(BTC)$_4$ treatment, Hes1 protein expression after 3 h was significantly lower than in the 30 min treatment group. This also shows that EGCG@MOF Zn(BTC)$_4$ can reduce the inflammation state of macrophages. Again, this was time-dependent.

Collectively, these results demonstrate that EGCG@MOF $Zn(BTC)_4$ can reduce the inflammatory response of macrophages caused by the Notch signaling pathway by reducing Notch protein secretion. This shows that EGCG@MOF $Zn(BTC)_4$ can facilitate the anti-inflammatory effect of EGCG.

After 3 h LPS (200 EU/mL) stimulation of mouse RAW264.7 macrophages, the cell supernatants of each group were collected, and the inflammatory factor TNF-α was detected. The results showed that TNF-α expression increased compared with the control group (Figure 4). After a 30 min treatment with EGCG@MOF $Zn(BTC)_4$, LPS-induced TNF-α secretion by macrophages was significantly reduced. This shows that EGCG@MOF $Zn(BTC)_4$ can effectively reduce the expression of inflammatory factors in mouse macrophages.

3.5. EGCG@MOF $Zn(BTC)_4$ Improved the Stability of EGCG Action

Establishing an animal model of diabetes is important for investigating diabetes and related complications. Common diabetes models include high-fat diet-induced, obesity-induced (ob/ob food-induced), genotype (db/db), and pharmacologically induced. Diabetes models successfully induced by drugs have the advantages of stability and fast modeling, so they are widely used in numerous studies of diabetes and its complications [29]. The ICR mouse is commonly used as an animal research object in the study of diabetes and its complications [30]. Here we injected ICR mice with STZ and performed continuous blood glucose testing to confirm the successful establishment of diabetic mice that were used in wound model experiments.

Studies have shown that EGCG can reduce inflammation in cells and tissues, suggesting that EGCG exerts anti-inflammatory effects [31]. In order to clarify whether EGCG@MOF $Zn(BTC)_4$ can improve EGCG stability and allow it to reduce inflammation, wounds were made on the dorsal skin of diabetic mice, and EGCG@MOF $Zn(BTC)_4$ or EGCG were applied once every four days. Images and measurements were taken daily to observe the effects of EGCG@MOF $Zn(BTC)_4$ and EGCG on wound healing in diabetic mice.

Wound healing evolution is shown in Figure 8A. To better explore the effects of EGCG@MOF $Zn(BTC)_4$, we generated a line chart of the wound non-healing rate.

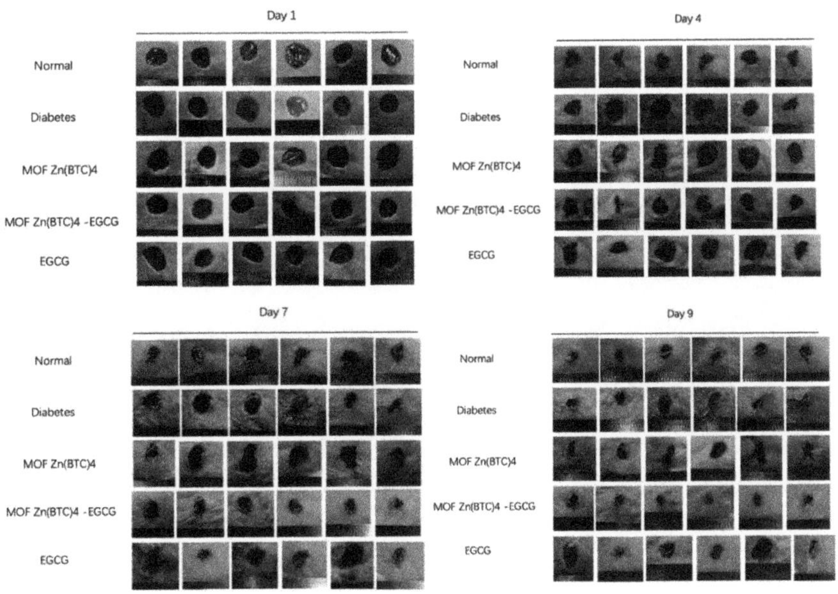

(A)

Figure 8. *Cont.*

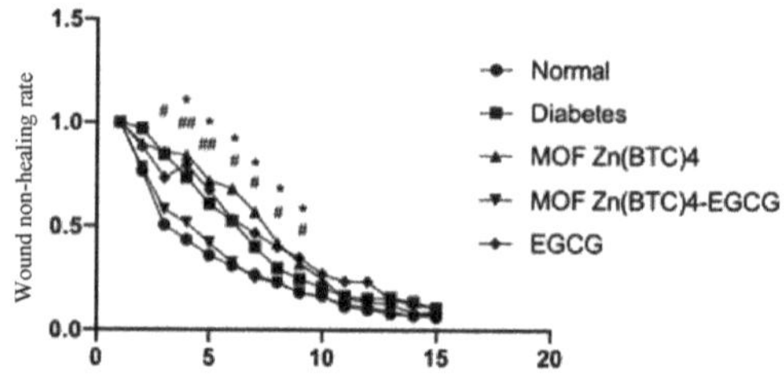

(**B**) Days of wound healing

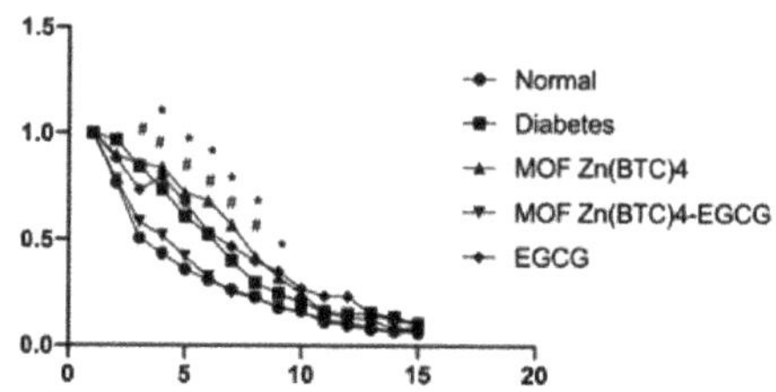

(**C**) Days of wound healing

Figure 8. Wound healing and the broken line chart of the wound non-healing rate. (**A**) Wound conditions at different time points with treatment every 4 days; (**B**) Wound non-healing rate in the normal and model groups; (**C**) Wound non-healing rate, normal group vs. MOF $Zn(BTC)_4$, $p < 0.05$; MOF $Zn(BTC)_4$ vs. EGCG@MOF $Zn(BTC)_4$, $p < 0.05$,B and C respectively indicate the arrogance between different groups.

Compared with the model group, normal mice showed significant differences on days three to nine (Figure 8B). At these time points, the proportions of unhealed wound area in normal mice were 50.5%, 43.41%, 35.96%, 31.05%, 26.56%, 23.01%, and 17.88%. The wound non-healing rates of diabetic mice were significantly higher than that of normal mice at 85.35%, 73.59%, 60.87%, 52.52%, 40%, 29.86%, and 24.7%. This shows that inducing experimental diabetes can indeed slow down the rate of wound healing.

EGCG@MOF $Zn(BTC)_4$ and EGCG were applied on the first day and every four days thereafter. Figure 8B shows that treatment groups showed significant differences on days four to nine. EGCG@MOF $Zn(BTC)_4$ more obviously improved wound healing in diabetic mice. This suggests that EGCG@MOF $Zn(BTC)_4$ can improve the stability of EGCG. At these time points, the non-healing rates of the EGCG@MOF $Zn(BTC)_4$ treatment group were 51.63%, 42.12%, 32.52%, 25.04%, 22.82%, and 18.06%, which were close to the healing rate of the normal group. The non-healing rates of the wounds in the EGCG-treated group were 79.27%, 67.62%, 53.27%, 46.67%, 39.92%, and 34.93%, which were similar to those of mice in the model group.

Compared with the Normal group, the MOF $Zn(BTC)_4$ group showed significant differences on days three to eight (Figure 8C). At these time points, the unhealed wound areas in MOF $Zn(BTC)_4$-treated mice were 85.38%, 72.08%, 68.37%, 56.94%, and 42.22%, respectively. The diabetic mouse wound model did not differ significantly from the model group after applying MOF $Zn(BTC)_4$, but the results were significantly different from those in the Normal group. This shows that MOF $Zn(BTC)_4$ alone does not significantly promote wound healing in mice. Compared with the EGCG@MOF $Zn(BTC)_4$ group, mice treated with MOF $Zn(BTC)_4$ also showed a significant difference on days three to eight. This indicates that it is the EGCG residing in EGCG@MOF $Zn(BTC)_4$ that promotes wound healing. The slow-release effect promotes anti-inflammatory effects and wound healing in diabetic mice. These results suggest that EGCG@MOF $Zn(BTC)_4$ improves the stability of EGCG.

3.6. EGCG@MOF $Zn(BTC)_4$ Promotes Skin Wound Healing in Diabetic Mice

Wound healing is a complex process that is influenced by many cytokines. Inflammation is the main factor that slows diabetic wound healing [29], which is also hindered by abnormal cellular responses. To investigate the effect of cytokines on wound healing in diabetic mice, we measured some of the cytokines released from the wound sites during the healing process. Researchers have shown that secretion of the inflammatory factor IL-1β is clearly related to slow wound healing. Reducing the release of IL-1β protein in tissue can accelerate the healing speed of diabetic wounds [31]. The pro-inflammatory factor IL-6 is also an important factor in this process.

Enzyme-linked immunosorbent assays were used to detect IL-1β and IL-6 in the wound skin tissue of diabetic mice on day nine (Figure 9).

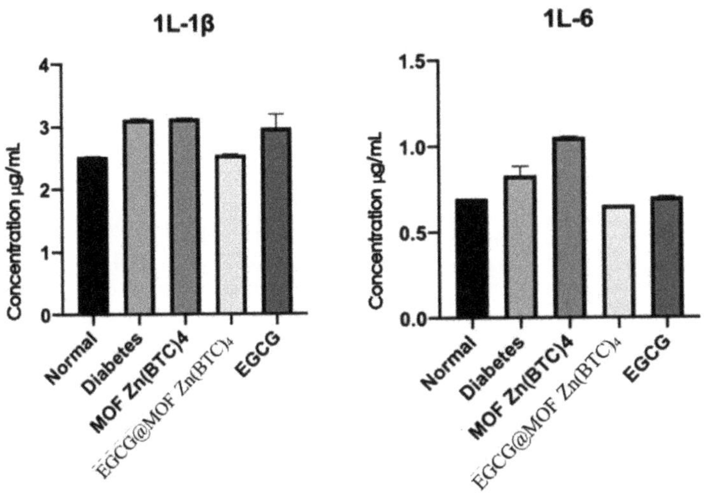

Figure 9. Expression of inflammatory factors IL-1β and IL-6 in wound tissue.

Compared with the normal group, levels of IL-1β released in the wounds of the diabetic model mice were increased, indicating a greater degree of inflammation. IL-1β levels in the MOF $Zn(BTC)_4$ and model groups were similar, suggesting that MOF $Zn(BTC)_4$ itself cannot produce an anti-inflammatory effect. After EGCG@MOF $Zn(BTC)_4$ treatment, the release of IL-1βwas significantly reduced in diabetic mice; it was almost as low as that measured in the normal group, and it was significantly lower than that of mice in the EGCG group. Again, this indicates that MOF $Zn(BTC)_4$ improves the stability of EGCG. Compared with normal skin tissue, the release level of the pro-inflammatory factor IL-6 was significantly increased in diabetic wound tissue, and the results were similar to those of IL-1β. Overall, wounds of diabetic mice treated with EGCG@MOF $Zn(BTC)_4$ showed obviously lower expression of inflammatory factors.

To further observe the effect of EGCG@MOF $Zn(BTC)_4$ on skin wound healing in diabetic mice, we performed hematoxylin and eosin (H&E) staining to detect re-epithelialization. To investigate if slow wound healing in diabetic mice was improved by EGCG@MOF $Zn(BTC)_4$ treatment on a microscopic level, skin wound tissues were collected on day nine for paraffin sectioning and conventional H&E. The differences in tissue re-epithelization are shown in Figure 10. The results indicate that EGCG@MOF $Zn(BTC)_4$ promotes wound healing in diabetic mice.

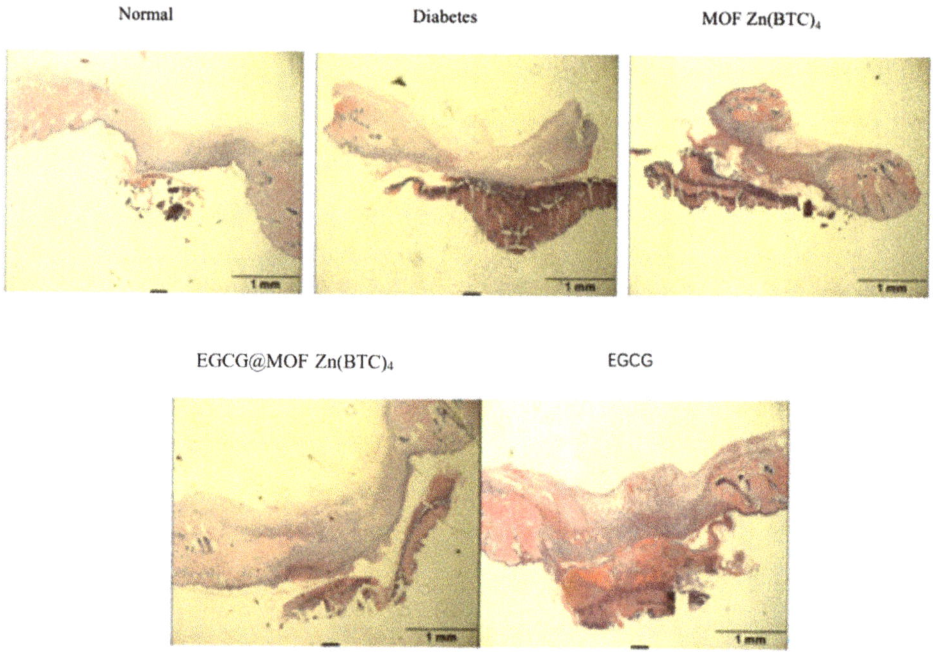

Figure 10. Tissue morphology in skin wounds.

The non-epithelialized area of skin tissue in the normal group was significantly reduced compared with the model group, and it was clearly observed whether there was a completely healed area in the model group of mice. The MOF $Zn(BTC)_4$ group did not have significant differences compared with the model group, and EGCG-treated mice did not show a significant improvement in re-epithelialization. The EGCG@MOF $Zn(BTC)_4$ treatment group and EGCG treatment. The mice in the group are close to the re-epithelialization of the skin tissue of the model group. This shows that the administration of EGCG alone every four days does not improve wound healing in diabetic mouse models.

Compared with the model and EGCG treatment groups, re-epithelialization was significantly improved in the EGCG@MOF $Zn(BTC)_4$ treatment group and was close to that of the normal group. Compared with the EGCG alone, EGCG@MOF $Zn(BTC)_4$ treatment significantly improves the re-epithelialization of skin wound tissue in diabetic mice. This again supports the hypothesis that EGCG@MOF $Zn(BTC)_4$ can improve EGCG stability.

Extracellular matrix (ECM) is a macromolecular substance synthesized by cells in the dermis. It is distributed between cells or on their surfaces, where it plays an indispensable role in wound healing. ECM components mainly include proteoglycans, proteins, and polysaccharides, and collagen is an important component and marker of ECM. The substances in ECM form a complex network structure that supports and connects the tissue structure, regulates tissue development, and impacts the physiological activities of cells. ECM formation is an important physiological reaction in the process of wound healing, and a higher level of ECM formation can promote skin wound healing. As chronic diabetic wounds heal, different types of infiltrating cells appear at the skin wound site, which reduces ECM formation and slows wound healing [32].

Researchers have shown that decreased collagen deposition will slow down the rate of wound healing [33]. To observe new ECM formation and assess the effect of EGCG@MOF $Zn(BTC)_4$ treatment, we used Masson's trichrome staining to observe collagen deposition levels on day nine of wound healing (Figure 11).

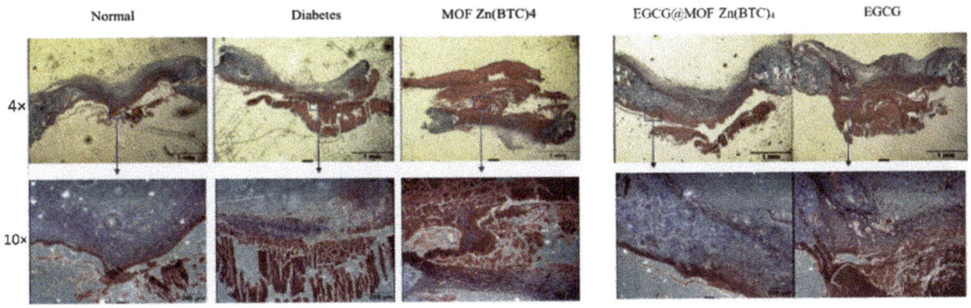

Figure 11. Collagen deposition.

Masson staining showed that collagen deposition (blue area) in the wounds of the diabetes model group was significantly lower than that of the normal group. Compared with the model group, the MOF Zn(BTC)$_4$ group did not show significant differences, and the mice in the EGCG treatment group did not show a significant improvement in collagen deposition. The mice in the MOF Zn(BTC)$_4$ and EGCG groups had collagen deposition similar to the model group. This suggests that EGCG administration every four days does not improve wound healing in a diabetic mouse model.

Compared with the model and EGCG treatments, collagen deposition was significantly improved in the EGCG@MOF Zn(BTC)$_4$ treatment group, with similar re-epithelialization as observed in the normal group. These results indicate that EGCG@MOF Zn(BTC)$_4$ can increase collagen deposition in skin wounds of diabetic mice compared with EGCG treatment alone, thus significantly improving ECM formation. This also indicates that EGCG@MOF Zn(BTC)$_4$ can improve the stability of EGCG for treatment purposes.

4. Conclusions

Our results demonstrate that MOF Zn(BTC)$_4$ can be loaded with EGCG. After optimizing the loading conditions, each gram of MOF Zn(BTC)$_4$ can be loaded with 236.227 mg EGCG. Chemical characterization of MOF Zn(BTC)$_4$ using IR spectroscopy showed that a new absorption peak appeared after adding EGCG. XRD analysis confirmed that the skeleton structure of MOF Zn(BTC)$_4$ filled with EGCG is basically maintained, and EGCG@MOF Zn(BTC)$_4$ can slowly release EGCG, which reduces the degradation rate of EGCG. By slowly releasing EGCG, EGCG@MOF Zn(BTC)$_4$ can reduce the expression of related inflammatory factors and significantly inhibit the inflammatory response in LPS-induced macrophages. In addition, EGCG@MOF Zn(BTC)$_4$ can significantly inhibit the expression of proteins related to the Notch signaling pathway and reduce the expression of inflammatory TNF-α. Wound healing experiments in diabetic mice revealed that adding EGCG to MOF Zn(BTC)$_4$ can slow down the degradation rate of EGCG, enabling slower release and prolonging the time for EGCG to function. EGCG@MOF Zn(BTC)$_4$ treatment can significantly reduce inflammation in the skin wound tissue of diabetic mice. Compared with the skin wound tissue of control diabetic mice, the re-epithelialization and collagen accumulation levels of the EGCG group were significantly increased. Collectively, the results demonstrate that EGCG@MOF Zn(BTC)$_4$ has the ability to release EGCG slowly, which slows its degradation and prolongs its action time.

Although MOFs have many significant advantages, there are still many challenges in clinical application. Compared with other nanocarriers, the development of MOFs as drug carriers is still in its infancy. Many problems need further optimization and research, such as the toxicity of metal ions, the degradability of MOFs, and the improvement of the biocompatibility of MOFs with the body. However, we have reason to believe that MOFs will have broad development prospects in the future.

Author Contributions: Data curation, J.Y., X.L. and S.L. (Senlin Li); Formal analysis, J.Y. and S.W.; Funding acquisition, S.L. (Song Li); Methodology, S.L. (Song Li) and J.S.; Project administration, J.Y.; Software, Q.Z. and X.W.; Writing—review & editing, S.L. (Song Li) and X.L. All authors have read and agreed to the published version of the manuscript.

Funding: The National Natural Science Foundation of China/Regional Science Foundation Project funded this research. Grant number No.21965038 and Applied Basic Research Program of Yunnan Province/General Projects of Basic Research. Grant number No.2019FB024.

Institutional Review Board Statement: Yunnan Agricultural University.

Informed Consent Statement: Not applicable.

Data Availability Statement: Not applicable.

Conflicts of Interest: The authors declare no conflict to interest.

Sample Availability: Samples of the compounds MOF $Zn(BTC)_4$ are available from the authors.

References

1. Toniolo, A.; Buccellati, C.; Pinna, C.; Gaion, R.M.; Sala, A.; Bolego, C. Cyclooxygenase-1 and Prostacyclin Production Endothelial Cells in the Presence of Mild Oxidative Stress. *PLoS ONE* **2013**, *8*, e56683. [CrossRef] [PubMed]
2. Amarajothi, D.; Li, Z.; Hermenegildo, G. Catalysis and photocatalysis by metal organic frameworks. *Chem. Soc. Rev.* **20** *47*, 8134–8172.
3. An, B.J.; Kwak, J.H.; Son, J.H.; Park, J.M.; Lee, J.Y.; Jo, C.; Byun, M.W. Biological and anti-microbial activity of irradiated green t polyphenols. *Food Chem.* **2004**, *88*, 549–555. [CrossRef]
4. Chen, J.; Abazari, R.; Adegoke, K.A.; Maxakato, N.W.; Bello, O.S.; Tahir, M.; Zhou, Y. Metal–organic frameworks and deriv materials as photocatalysts for water splitting and carbon dioxide reduction. *Coord. Chem. Rev.* **2022**, *469*, 214664. [CrossRef]
5. Artavanistsakonas, S.; Delidakis, C.; Fehon, R.G. The Notch Locus and the Cell Biology of Neuroblast Segregation. *Annu. R Cell Biol.* **1991**, *7*, 427–452. [CrossRef] [PubMed]
6. Cabrera, C.; Artacho, R.; Giménez, R. Beneficial effects of green tea—A review. *J. Am. Coll. Nutr.* **2006**, *25*, 79–99. [CrossRef] [PubMe
7. Abazari, R.; Amani-Ghadim, A.R.; Slawin, A.M.; Carpenter-Warren, C.L.; Kirillov, A.M. Non-Calcined Layer-Pillare Mn0.5Zn0.5 Bimetallic–Organic Framework as a Promising Electrocatalyst for Oxygen Evolution Reaction. *Inorg. Chem.* **202** *61*, 9514–9522. [CrossRef]
8. Cabrera, C.; Giménez, R.; López, M. Determination of Tea Components with Antioxidant Activity. *J. Agric. Food Chem.* **2003**, 5 4427–4435. [CrossRef]
9. Pan, Y.; Abazari, R.; Yao, J.; Gao, J. Recent progress in 2D metal-organic framework photocatalysts: Synthesis, photocatalyt mechanism and applications. *J. Phys. Energy* **2021**, *3*, 032010. [CrossRef]
10. Chakrawarti, L.; Agrawal, R.; Dang, S.; Gupta, S.; Gabrani, R. Therapeutic effects of EGCG: A patent review. *Expert Opin. Ther. P* **2016**, *26*, 907–916. [CrossRef]
11. Coussens, L.M.; Werb, Z. Inflammation and cancer. *Nature* **2002**, *420*, 860–867. [CrossRef] [PubMed]
12. Luo, X.; Abazari, R.; Tahir, M.; Fan, W.K.; Kumar, A.; Kalhorizadeh, T.; Kirillov, A.M.; Amani-Ghadim, A.R.; Chen, J.; Zhou, Trimetallic metal–organic frameworks and derived materials for environmental remediation and electrochemical energy storag and conversion. *Coord. Chem. Rev.* **2022**, *461*, 214505. [CrossRef]
13. Dekel, Y.; Glucksam, Y.; Elron-Gross, I.; Margalit, R. Insights into modeling streptozotocin-induced diabetes in ICR mice. *Lab Anir* **2009**, *38*, 55. [CrossRef] [PubMed]
14. Giniger, E. Notch signaling and neural connectivity. *Curr. Opin. Genet. Dev.* **2012**, *22*, 339–346. [CrossRef]
15. Kalhorizadeh, T.; Dahrazma, B.; Zarghami, R.; Mirzababaei, S.; Kirillov, A.M.; Abazari, R. Quick removal of metronidazole fror aqueous solutions using metal–organic frameworks. *New J. Chem.* **2022**, *46*, 9440–9450. [CrossRef]
16. Abazari, R.; Sanati, S.; Morsali, A. Mixed Metal Fe_2Ni MIL-88B Metal–Organic Frameworks Decorated on Reduced Graphen Oxide as a Robust and Highly Efficient Electrocatalyst for Alkaline Water Oxidation. *Inorg. Chem.* **2022**, *61*, 3396–3405. [CrossRe
17. Horcajada, P.; Serre, C.; Vallet-Regí, M.; Sebban, M.; Taulelle, F.; Férey, G. Metal–Organic Frameworks as Efficient Materials fo Drug Delivery. *Angew. Chem. Int. Ed.* **2006**, *118*, 6120–6124. [CrossRef]
18. Xie, H.; Sun, J.; Chen, Y.; Zong, M.; Li, S.; Wang, Y. EGCG Attenuates Uric Acid-Induced Inflammatory and Oxidative Stres Responses by Medicating the NOTCH Pathway. *Oxidative Med. Cell. Longev.* **2015**, *2015*, 1–10. [CrossRef]
19. Kopan, R.; Ilagan, M. The Canonical Notch Signaling Pathway: Unfolding the Activation Mechanism. *Cell* **2009**, *13* 216–233. [CrossRef]
20. Li, B.; Wen, H.-M.; Cui, Y.; Zhou, W.; Qian, G.; Chen, B. Emerging Multifunctional Metal–Organic Framework Material *Adv. Mater.* **2016**, *28*, 8819–8860. [CrossRef]
21. Li, W.-J.; Tu, M.; Cao, R.; Fischer, R.A. Metal–organic framework thin films: Electrochemical fabrication techniques an corresponding applications & perspectives. *J. Mater. Chem. A* **2016**, *4*, 12356–12369.
22. Mendez, J.D.; Ramos, H.G. Animal Models in Diabetes Research. *Arch. Med. Res.* **1994**, *25*, 367. [PubMed]

Mirza, R.E.; Fang, M.M.; Ennis, W.J.; Koh, T.J. Blocking interleukin-1β induces a healing-associated wound macrophage phenotype and improves healing in type 2 diabetes. *Diabetes* **2013**, *62*, 2579–2587. [CrossRef] [PubMed]

Abu-Dief, A.M.; Alrashedee, F.M.; Emran, K.M.; Al-Abdulkarim, H.A. Development of some magnetic metal–organic framework nano composites for pharmaceutical applications. *Inorg. Chem. Commun.* **2022**, *138*, 109251. [CrossRef]

5. Morain, W.D.; Colen, L.B. Wound healing in diabetes mellitus. *Clin. Plast. Surg.* **1990**, *17*, 493. [CrossRef]

6. Pernot, L.; Frénois, F.; Rybkine, T.; L'Hermite, G.; Petrella, S.; Delettré, J.; Sougakoff, W. Crystal structures of the class D beta-lactamase OXA-13 in the native form and in complex with meropenem. *J. Mol. Biol.* **2001**, *310*, 859–874. [CrossRef]

7. Portera, C.A.; Love, E.J.; Memore, L.; Zhang, L.; Müller, A.; Browder, W.; Williams, D.L. Effect of macrophage stimulation on collagen biosynthesis in the healing wound. *Am. Surg.* **1997**, *63*, 125–131.

8. Pullikotil, P.; Chen, H.; Muniyappa, R.; Greenberg, C.C.; Yang, S.; Reiter, C.E.; Lee, J.-W.; Chung, J.H.; Quon, M.J. Epigallocatechin gallate induces expression of heme oxygenase-1 in endothelial cells via p38 MAPK and Nrf-2 that suppresses proinflammatory actions of TNF-α. *J. Nutr. Biochem.* **2012**, *23*, 1134–1145. [CrossRef]

9. Syed, D.N.; Afaq, F.; Kweon, M.H.; Hadi, N.; Bhatia, N.; Spiegelman, V.S.; Mukhtar, H. Green tea polyphenol EGCG suppresses cigarette smoke condensate-induced NF-kappaB activation in normal human bronchial epithelial cells. *Oncogene* **2007**, *26*, 673–682. [CrossRef]

0. Wellen, K.E.; Hotamisligil, G.S. Inflammation, stress, and diabetes. *J. Clin. Investig.* **2005**. [CrossRef]

1. Zhang, W.; Xu, W.; Xiong, S. Blockade of Notch1 Signaling Alleviates Murine Lupus via Blunting Macrophage Activation and M2b Polarization. *J. Immunol.* **2010**, *184*, 6465–6478. [CrossRef] [PubMed]

2. Zhou, D.; Huang, C.; Lin, Z.; Zhan, S.; Kong, L.; Fang, C.; Li, J. Macrophage polarization and function with emphasis on the evolving roles of coordinated regulation of cellular signaling pathways. *Cell. Signal.* **2014**, *26*, 192–197. [CrossRef] [PubMed]

3. Zscheck, K.K.; Murray, B.E. Genes involved in the regulation of beta-lactamase production in enterococci and staphylococci. *Antimicrob. Agents Chemother.* **1993**, *37*, 1966–1970. [CrossRef] [PubMed]

molecules

Article

Cooling Capacity Test for MIL-101(Cr)/$CaCl_2$ for Adsorption Refrigeration System

Zhongbao Liu [1,*], Banghua Zhao [1], Yong Huang [2], Xin Qi [3] and Fengfei Lou [1]

1 Department of Refrigeration and Cryogenic Engineering, College of Environmental and Energy Engineering, Beijing University of Technology, 100 Pingleyuan Road, Chaoyang, Beijing 100124, China; zhaobanghua@emails.bjut.edu.cn (B.Z.); loufengfei@emails.bjut.edu.cn (F.L.)
2 Suzhou Sujing Anfa Air Conditioning Co. LTD, 2 Weixin Road, Suzhou Industrial Park, Suzhou 215112, China; HY@aimfar.com.cn
3 China Household Electric Appliance Research Institute, 6 Yuetan beixiao Str, Xicheng, Beijing 100037, China; qix@cheari.com
* Correspondence: liuzhongbao@bjut.edu.cn; Tel./Fax: +86-010-67391613

Academic Editors: Victoria Samanidou and Eleni Deliyanni
Received: 7 August 2020; Accepted: 26 August 2020; Published: 31 August 2020

Abstract: An MIL-101(Cr) powder material was successfully prepared using the hydrothermal synthesis method, and then the original MIL-101(Cr) was combined with different mass fractions of $CaCl_2$ using the immersion method to obtain a MIL-101(Cr)/$CaCl_2$ composite material. The physical properties of the adsorbent were determined by X-ray powder diffraction (XRD), an N2 adsorption desorption isotherm test, and thermogravimetric analysis (TG). The water vapor adsorption performance of the metal-organic frameworks MOFs was tested with a gravimetric water vapor adsorption instrument to analyze its water vapor adsorption mechanism. Based on the SIMULINK platform in the MATLAB software, a simulation model of the coefficient of performance (COP) and cooling capacity of the adsorption refrigeration system was established, and the variation trends of the COP and cooling capacity of the adsorption refrigeration system under different evaporation/condensation/adsorption/desorption temperatures was theoretically studied. MIL101-(Cr)/$CaCl_2$-20% was selected as the adsorption material in the adsorption refrigeration system through the physical characterization of composite materials with different $CaCl_2$ concentrations by means of adsorption water vapor test experiments. A closed adsorption system performance test device was built based on the liquid level method. The cooling power per unit and adsorbent mass (COP and SCP) of the system were tested at different evaporation temperatures (288 K/293 K/298 K); the adsorption temperature was 298 K, the condensation temperature was 308 K, and the desorption temperature was 353 K. The experimental results showed that COP and SCP increased with the increase in the evaporation temperature. When the evaporation temperature was 298 K, the level of COP was 0.172, and the level of SCP was 136.9 W/kg. The COP and SCP of the system were tested at different adsorption temperatures (293 K/298 K/303 K); the evaporation temperature was 288 K, the condensation temperature was 308 K, and the desorption temperature was 353 K. The experimental results showed that the levels of COP and SCP decreased with the increase in the adsorption temperature. When the adsorption temperature was 293 K, the level of COP was 0.18, and the level of SCP was 142.4 W/kg.

Keywords: MIL-101(Cr)/$CaCl_2$-20%; composite materials; modeling and simulation; adsorption performance

1. Introduction

Adsorption refrigeration is an environmentally friendly refrigeration method. Adsorption refrigerators can use natural working fluids as refrigerants and have zero Ozone Depletion Potential (ODP), which has attracted widespread attention in recent years. Low-temperature heat-driven environmentally friendly adsorption refrigeration systems are becoming a viable alternative to electric-driven vapor compression refrigeration systems. The low cooling capacity of the technology hinders its successful commercialization [1–3].

At present, there are two main ways to improve the performance of an adsorption refrigeration system. One is to optimize the circulation of the system, and the other is to improve the adsorption performance of the working medium [4]. Metal-organic frameworks (MOFs) are a new type of solid adsorbent with a high specific surface area (4000 $m^2 \cdot g^{-1}$) and porosity, which have broad application prospects in TDCs or adsorption heat pumps (AHPs) [5]. The following is a review of the literature on adsorption refrigeration studies using metal-organic frameworks (MOFs) as adsorbents. Xia [6] synthesized three Zr-MOFs with different pore sizes: UiO-66, UiO-67 and NU-1000, and determined their adsorption isotherms to water and ethanol. The cooling power per unit (COP) of MOF/water and MOF/ethanol were studied by adsorption system model. The results showed that the COP of MOF/water was generally higher than that of MOF/ethanol, but the performance of MOF/ethanol was better than that of MOF/water. Ehrenman [7] tested the water absorption capacity and cyclic stability of the high-porosity metallic organic framework material MIL-101 (3D-[$Cr_3F(H_2O)_2O(bdc)_3 \cdot 25H_2O$]) at 25 °C. An N_2 adsorption isotherm showed that, after 20 adsorption and desorption cycles, the specific surface area of the Brunauer-Emmett-Teller (BET) of MIL-101(Cr) decreased from 2059 m^2/g to 2047 m^2/g, and its water absorption capacity decreased by 1.9%. Ahmed Rezk [8] studied the water absorption properties of six different MOF materials (Cu-BTC, Fe-BTC, MIL-53, Birm-1, Birm-1-K and Birm-1-Li). The results show that the surface areas of Cu-BTC and Fe-BTC materials are 2100 m^2/g and 1600 m^2/g, respectively, which have the highest water absorption performance. Ahmed Rezk [9] also reported the isothermal adsorption properties of six different MOF materials (MIL-101(Cr), MIL-100 (Cr), MIL-53 (Cr), CPO-27-Ni, Cu-BTC, Fe-BTC) and silica gel to ethanol. The adsorption isotherms of various adsorbents were measured at 25 °C; the results showed that MIL-101(Cr) was superior to other adsorbents in ethanol adsorption. The stability of MIL-101(Cr)/ethanol in 20 consecutive adsorption–desorption cycles was also studied; the results show that the working pair has good stability. Baosheng Shi [10] compared the adsorption properties of silica gels RD-2060, SAPO-34 and CPO-27 (Ni). Under the condition of 0.1 relative humidity, the water vapor absorption rate of CPO-27 (Ni) was 0.41 kg/kg, 86.3% higher than SAPO-34, and 7 times higher than RD-2060. Teo [11] modified MIL-101(Cr) with 5%, 20% and 40% alkali metal ions (Li^+, Na^+, K^+), respectively, to test the properties of MOF materials after modification. Garzon-tovar [12] synthesized a new solid composite adsorbent by combining metal organic framework UiO-66 and hygroscopic salt $CaCl_2$ through spray drying, improving the water absorption capacity of the original material. At the same time, composite adsorption and materials were applied in an adsorption refrigeration system, and the results showed that the SCP was 631 W/kg and the COP was 0.83. Jian Yan [13] developed a new composite material (MIL-101@GO) for adsorption heat pump (AHPs). The steam adsorption isotherm and adsorption kinetics of the composite materials were determined by the weight method. The results showed that MIL-101@GO has a super high water vapor adsorption capacity, and its adsorption capacity can reach 1.58 g/g. Six consecutive adsorption cycle experiments showed that the synthesized MIL-101@GO has good reversibility and stability for water vapor adsorption.

Calcium chloride is a very hygroscopic salt, it has a strong affinity for water. It has been reported in the literature that when salt is combined with activated carbon, silica gel, zeolite and other porous materials (matrix) to form a composite material, its performance is significantly improved. Aristov [14] discusses a silica gel/$CaCl_2$ composite material called a selective water absorber (SWS-1L) in which the enhanced water vapor capacity increases in COP from 0.48 to 0.7 at the desorption temperature of 90 °C, evaporation temperature of 5 °C, and condensation temperature and adsorption temperature of

40 °C. Wang [15] combined activated carbon fiber felts (ACF FELT) with 30% calcium chloride, and the water adsorption capacity reached 1.7 g^{H2O}/g^{ads}, higher than activated carbon, silica gel and even the compound of silica gel and calcium chloride. Chan [16] soaked zeolite 13X in 46% $CaCl_2$ solution, and the adsorption capacity increased by 0.4 g^{H2O}/g^{ads} at 25 and 75 °C and 870 Pa.

In this paper, MIL-101(Cr)/$CaCl_2$ composite materials with different $CaCl_2$ concentrations (10%, 20%, 30%) were prepared by the dipping method. Through physical characterization and water vapor adsorption performance experiments, MIL-101(Cr)/$CaCl_2$-20%/water was selected as the pair of working media for the adsorption refrigeration system to test the system performance at different evaporation/adsorption temperatures.

2. Results

2.1. XRD Analysis

Figure 1 shows the XRD spectra of MIL-101(Cr)/$CaCl_2$ composites. It can be seen from the figure that the main characteristic diffraction peaks of MIL-101(Cr) samples appear at $2\theta = 3.42°, 5.98°, 8.56°, 9.18°, 10.30°$ and $16.54°$, and the position of the diffraction peak is very consistent with the report [15]. The MIL-101(Cr)/$CaCl_2$ composite has the same diffraction peak as the original MIL-101(Cr), indicating that MIL-101(Cr) is still the main component of the MIL-101(Cr)/$CaCl_2$ composite. In addition, we can also find that the peak of XRD diffraction gradually decreases with the increase in $CaCl_2$ content, because the disordered hydrated salt molecules are filled into the mesopores of MIL-101(Cr) with the increase in $CaCl_2$ content [17].

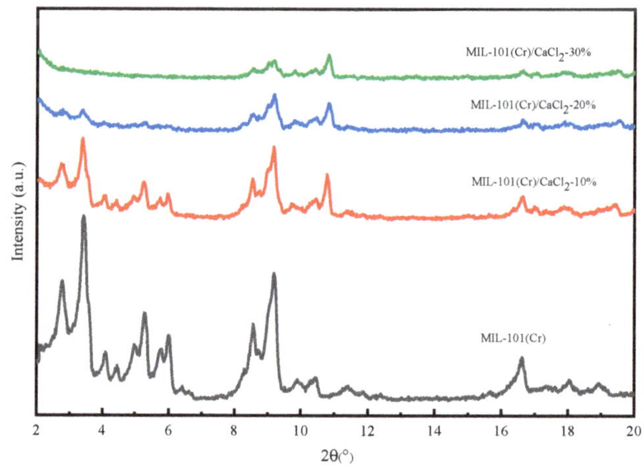

Figure 1. XRD pattern of MIL-101(Cr) and composites.

2.2. TG Analysis

The thermogravimetric results for MIL-101(Cr)/$CaCl_2$ composites are shown in Figure 2. The weightlessness of the material mainly goes through two stages. As the temperature rises to 100 °C, the water in MIL-101(Cr) dissipates, and the weight loss in the first stage is caused by the loss of calcium chloride in the composite. At 300 °C, MIL-101(Cr) began to lose the crystal water in the structure, and by 400 °C, the framework was decomposed. The structure of the composite material was destroyed at 450 °C, which was delayed by 50 °C compared with MIL-101(Cr).

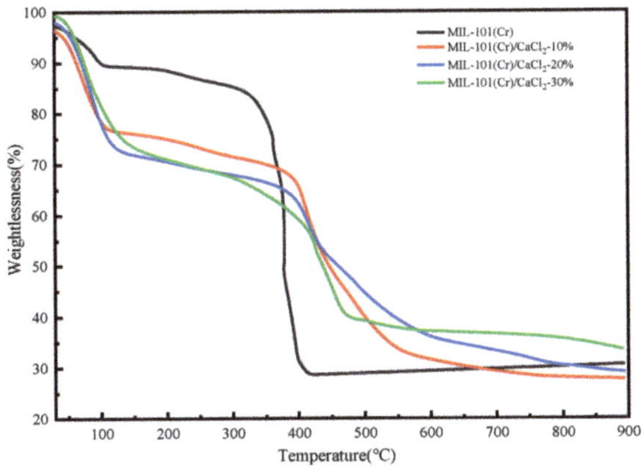

Figure 2. Thermogravimetric TG curve of MIL-101(Cr) and composites.

2.3. N_2 Adsorption–Desorption Isotherms and Pore Size Analysis

Figure 3 shows the N_2 adsorption and desorption isotherms for the MIL-101(Cr)/$CaCl_2$ composites. According to the classification of the BET adsorption isotherm curve and type-II adsorption–desorption isotherms, in the low pressure stage (p = 0–0.2/p_0), the nitrogen adsorption quantity displays sharp growth, proving that it contains a certain amount of microporous materials. At a relative pressure of about 0.2, the material adsorbed a second time, indicating that the experimentally synthesized material contained two different cage structures [18]. The adsorption capacity in the middle section of the curve (p/p_0 = 0.2–0.9) increased slowly with the increase in pressure. In the second half of the curve (p/p_0 = 0.9–1.0), the adsorption line rose sharply and did not show adsorption saturation until the vapor pressure was close to saturation, indicating that the sample contained a certain amount of mesoporous and macroporous bulk filling due to capillary condensation.

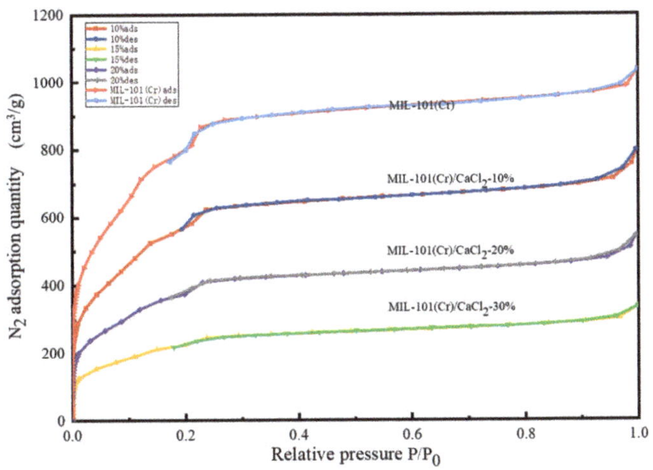

Figure 3. Nitrogen adsorption–desorption isotherms of MIL-101(Cr) and composites.

Figure 4 shows the pore size distribution of MIL-101(Cr) and MIL-101(Cr)/$CaCl_2$ composites. It can be seen from the figure that the MIL-101(Cr)/$CaCl_2$ composite pore diameters are mainly distributed at 0.86 nm, 1.19 nm and 2.32 nm. Table 1 lists the pore structure parameters of the MIL-101(Cr) and MIL-101(Cr)/$CaCl_2$ composites. With the increase in $CaCl_2$ addition, the specific surface area, total pore volume and pore size distribution of microporous mesopores of MIL-101(Cr)/$CaCl_2$ composites decreased gradually. This is because the added $CaCl_2$ molecules occupy part of the pores of the MIL-101(Cr) material. For MIL-101(Cr)/$CaCl_2$-30%, the BET specific surface area and total pore volume were only 193 m^2/g and 0.071 cm^3/g, which indicates that the $CaCl_2$ molecule almost completely blocked the pores of MIL-101(Cr).

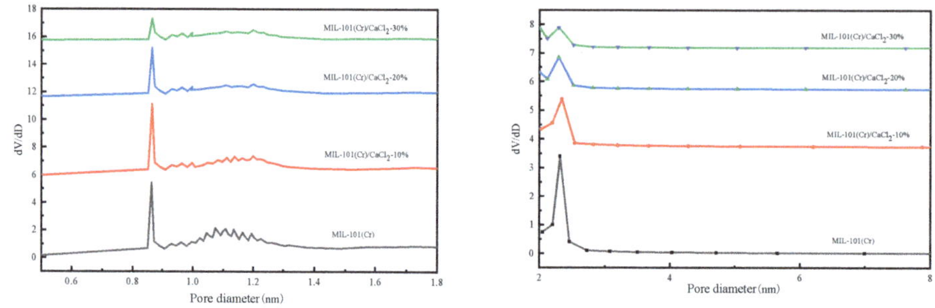

(**a**) (**b**)

Figure 4. Pore diameter distribution of MIL-101(Cr) and MIL-101(Cr)/$CaCl_2$ composites: (**a**) micropore, (**b**) mesoporous.

Table 1. Pore structure parameters of MIL-101(Cr) and composite materials.

Sample	BET Specific Surface Area (m^2/g)	Pore Diameter (cm^3/g)
MIL-101(Cr)	2824	1.362
MIL-101(Cr)/$CaCl_2$-10%	1977	0.745
MIL-101(Cr)/$CaCl_2$-20%	1307	0.509
MIL-101(Cr)/$CaCl_2$-30%	193	0.071

2.4. Water Vapor Adsorption Isotherms

It can be seen from Figure 5 that the MIL-101(Cr)/$CaCl_2$-10% adsorption isotherm line type is almost the same as the MIL-101(Cr) line type. The adsorption isotherms of MIL-101(Cr)/$CaCl_2$-20% and MIL-101(Cr)/$CaCl_2$-30% showed obvious changes, and the equilibrium adsorption capacity was higher than that of the original MIL-101(Cr). As can be seen from the adsorption isotherm, water absorbing capacity of calcium chloride itself leads to the low pressure area which in the water vapor adsorption experiment showed higher adsorption performance, the relative pressure adsorption amount compared with the original start going down after 0.4. This is due to the good water absorption characteristics of calcium chloride. The specific surface area of the material gradually became dominant; the specific surface area of the composites compared with MIL-101 decreased, so the relative pressure was greater than 0.4 after the adsorption performance was reduced. In particular, the MIL-101(Cr)/$CaCl_2$-20% and MIL-101(Cr)/$CaCl_2$-30% composite materials had a higher water adsorption capacity and equilibrium adsorption capacity than the original MIL-101(Cr) at low pressure. Therefore, the MIL-101(Cr)/$CaCl_2$-20% and MIL-101(Cr)/$CaCl_2$-30% materials are very attractive adsorbent materials for use in adsorption refrigeration applications. Although the equilibrium water adsorption capacity of the MIL-101(Cr)/$CaCl_2$-30% composite adsorbent is high, it becomes agglomerated after water absorption continues to agglomerate after repeated drying, which significantly affects the material's adsorption capacity and limits its use in adsorption refrigeration applications. In summary, the MIL-101(Cr)/$CaCl_2$-20% composite material was selected for its adsorption refrigeration performance.

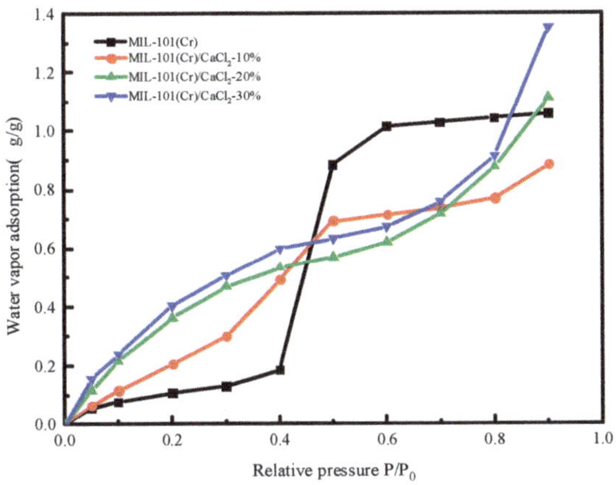

Figure 5. Water sorption isotherms of MIL-101(Cr) andMIL-101(Cr)/$CaCl_2$ composites at 298 K.

2.5. Analysis of the Simulation Results

2.5.1. Effect of Different Desorption Temperatures on System Cooling Capacity and COP

The effect of the desorption temperature T_{g2} on cooling capacity and COP is shown in Figure 6. It can be seen from the figure that the cooling capacity of the system increased with the increase in the desorption temperature. The total cycle time was 9000 s. From the COP, there was an optimal desorption temperature. When $T_e = 288$ K, $T_c = 308$ K, $Ta_2 = 298$ K, the optimal desorption temperature was 351 K. At this time, the system cooling capacity was 118 kJ and the COP was 0.24. Before the optimum desorption temperature, the COP value increased with the desorption temperature in a certain temperature range. The higher the desorption temperature, the more water vapor is desorbed from the adsorption bed, so the more water vapor is adsorbed by the material. The more the corresponding cooling capacity, the greater the amount of cooling. However, when the desorption temperature exceeds the optimum desorption temperature, the COP decreases as the desorption temperature increases because the desorption temperature increases and the heat consumed by the system cycle also increases, causing the COP to decrease.

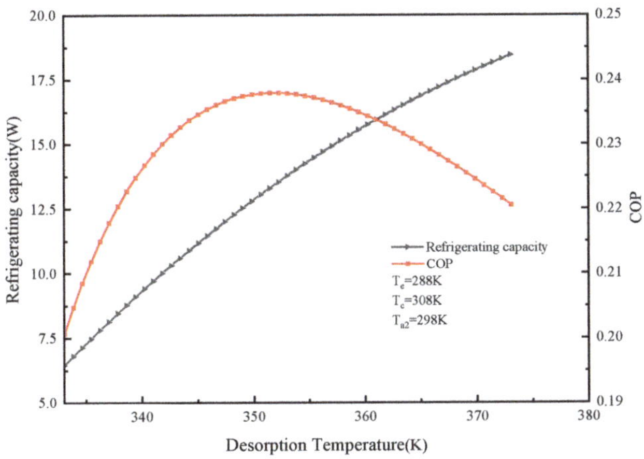

Figure 6. Effect of desorption temperature on refrigeration capacity and coefficient of performance (COP).

2.5.2. Effect of Different Condensation Temperatures on System Cooling Capacity and COP

Calculation parameters: T_e = 288 K, T_{a2} = 298 K, T_{g2} = 353 K. The effect of condensation temperature Tc on cooling capacity and COP is shown in Figure 7. It can be seen from the figure that as the condensing temperature increases, the cooling capacity and COP of the system decrease. When T_c = 313 K, the cooling capacity was 10.04 W and the COP was 0.17. This is because the condensation temperature is not conducive to the condensation of the refrigerant vapor, reducing the desorption amount of the refrigerant, thereby reducing the COP of the system. At the same time, the adsorption delay occurs when the condensation temperature rises, so that the adsorption amount decreases, causing the cooling capacity to decrease accordingly. It can be seen from the above law that reducing the condensation temperature of the adsorption refrigeration system can improve the performance of the system.

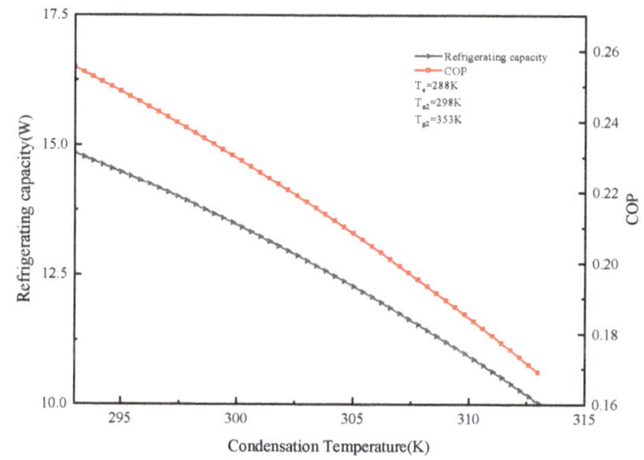

Figure 7. Effect of condensation temperature on refrigeration capacity and COP.

2.5.3. Effects of Different Adsorption Temperatures on Cooling Capacity and COP of the System

Calculation parameters: T_e = 288 K, T_c = 308 K, T_{g2} = 353 K. The effect of condensation temperature T_{a2} on cooling capacity and COP is shown in Figure 8. It can be seen from the figure that as the adsorption temperature increases, the cooling capacity and COP of the system decrease. The adsorption temperature increases, so that the difference between the adsorption temperature and the initial adsorption temperature decreases, and then the cyclic adsorption amount also decreases, eventually leading to a decrease in system COP. When T_e = 288 K, T_c = 308 K, T_{g2} = 353 K, and T_{a2} = 308 K, the minimum cooling capacity and COP were 8.81 W and 0.15, respectively.

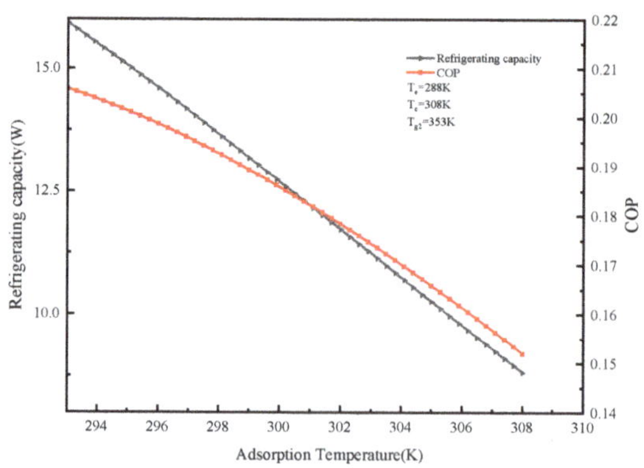

Figure 8. Effect of adsorption temperature on refrigeration capacity and COP.

2.5.4. Effects of Different Evaporation Temperatures on Cooling Capacity and COP of the System

Calculation parameters: T_{a2} = 298 K, T_c = 308 K, T_{g2} = 353 K. The effect of evaporation temperature (T_e) on the cooling capacity and COP is shown in Figure 9. It can be seen from the figure that the system's cooling capacity and coefficient of performance COP both increase with the increase in the evaporation temperature. When the evaporation temperature increased from 288 K to 303 K, the cooling capacity and COP—initially 8.81 W and 0.15—increased to 16 W and 0.28. This occurs because (1) the higher the evaporation temperature, the higher the evaporation pressure; (2) the larger the pressure difference in the adsorption bed, the smaller the mass transfer resistance; (3) the larger the adsorption amount of the adsorbent, the larger the sensible heat of the adsorption bed. Ultimately, the cooling capacity and coefficient of COP performance both increase as the evaporation temperature increases.

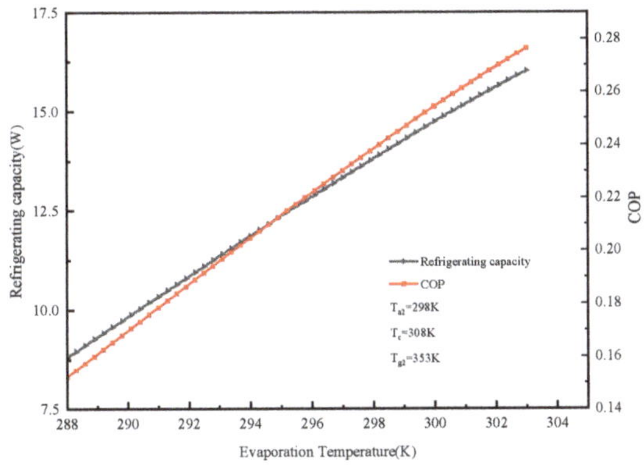

Figure 9. Effect of evaporation temperature on refrigeration capacity and COP.

2.6. Analysis of the Experimental Results

2.6.1. System Performance Analysis at Different Evaporation Temperatures

Figure 10 shows the water vapor adsorption capacity of the adsorption refrigeration system over time at different evaporation temperatures. It can be seen from the figure that when T_{a2} = 298 K, T_c = 308 K, T_{g2} = 353 K, the evaporation temperature was 288 K, the adsorption capacity was 0.625 g/g; when the evaporation temperature was 293 K, the adsorption capacity was 0.715 g/g; when the evaporation temperature was 298 K, the adsorption capacity was 0.845 g/g. We can find that as the evaporation temperature increases, the equilibrium adsorption amount of water vapor temperature also increases. This is because the evaporation pressure increases and the pressure difference in the refrigeration system also increases, which is beneficial to MIL-101(Cr)/$CaCl_2$-20% composite adsorption.

It can be seen from Figure 11 that as the evaporation temperature increases, the cooling capacity increases. When the evaporation temperature was 298 K, the cooling capacity of the system increased by 5.47 W compared with that of the system when the evaporation temperature was 288 K.

During the experiment, as the evaporation temperature increases, the system cooling capacity increases, resulting in an upward trend in the system's COP and SCP. It can be found from Figure 12 that the COP of the system was 0.103 when the evaporation temperature was 288 K. When the evaporation temperature rose to 298 K, the COP of the system was 0.172, which was increased by 67%. For SCP, when the evaporation temperature was 288 K, the value was 82.2 W/kg, and when the evaporation temperature was 298 K, the SCP value was 136.9 W/kg, which is an increase of nearly 66.5%. Therefore, in the refrigeration process, in the case of meeting the cooling demand, the evaporation temperature of the system should be increased as much as possible.

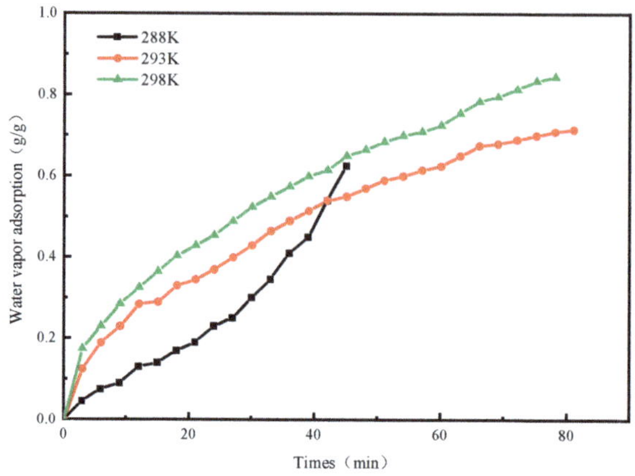

Figure 10. The curve of adsorption uptake of water vapor versus time.

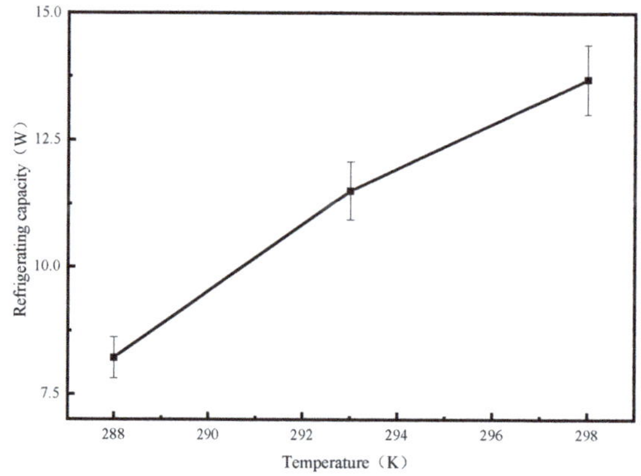

Figure 11. Cooling capacity of the system at different evaporation temperatures.

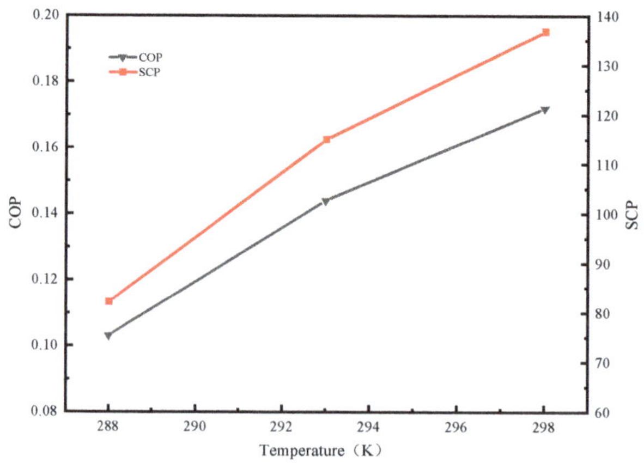

Figure 12. COP and adsorbent mass (SCP) of the system at different evaporation temperatures.

2.6.2. System Performance Analysis at Different Adsorption Temperatures

Figure 13 shows the variation of system adsorption over time at different adsorption temperatures. It can be seen from the figure that when T_e = 288 K, T_c = 308 K, T_{g2} = 353 K, the adsorption temperature was 293 K, the adsorption capacity was 0.9 g/g; when the adsorption temperature was 298 K, the adsorption capacity was 0.66 g/g; when the adsorption temperature was 303 K, the adsorption capacity was 0.565 g/g. From this we can see that as the adsorption temperature decreases, the amount of water vapor adsorption in the system gradually increases. The increase in the adsorption temperature causes the relative pressure to decrease. It can be seen from the adsorption isotherm of Figure 5 that as the relative pressure decreases, the equilibrium adsorption capacity decreases.

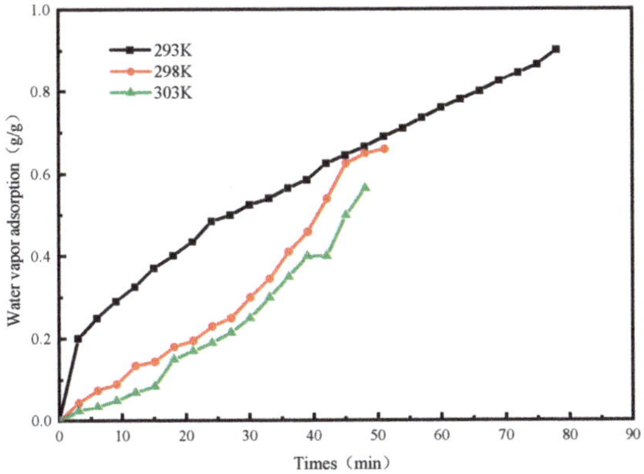

Figure 13. The curve of adsorption uptake of water vapor versus time.

It can be seen from Figure 14 that the cooling capacity decreases with increasing adsorption temperature. When the adsorption temperature was 293 K, the system cooling capacity increased by 4.94W compared with that of when the adsorption temperature was 303 K.

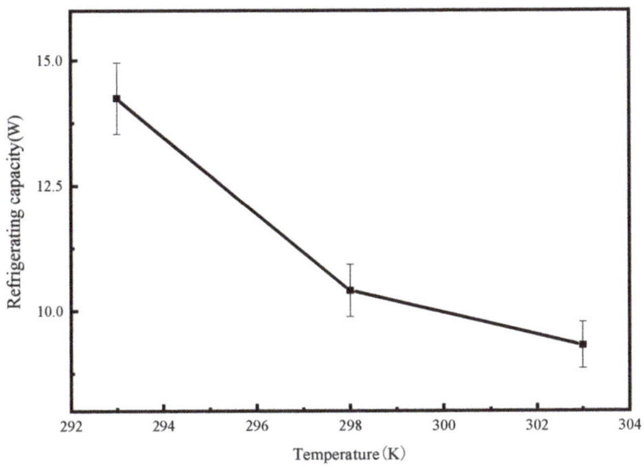

Figure 14. Cooling capacity of the system at different adsorption temperatures.

It can be seen from Figure 15 that the COP and SCP of the system decrease with the increase in adsorption temperature, and the downward trend is very obvious. When the adsorption temperature was 293 K, the COP and SCP of the system were 0.18 and 142.4 W/kg, respectively. When the adsorption temperature was raised to 303 K, the COP and SCP of the system were reduced to 0.12 and 93 W/kg, respectively, and the performance of the system was nearly doubled compared with the adsorption temperature of 293 K. Therefore, in the adsorption refrigeration system, the adsorption temperature has a significant influence on the performance of the system.

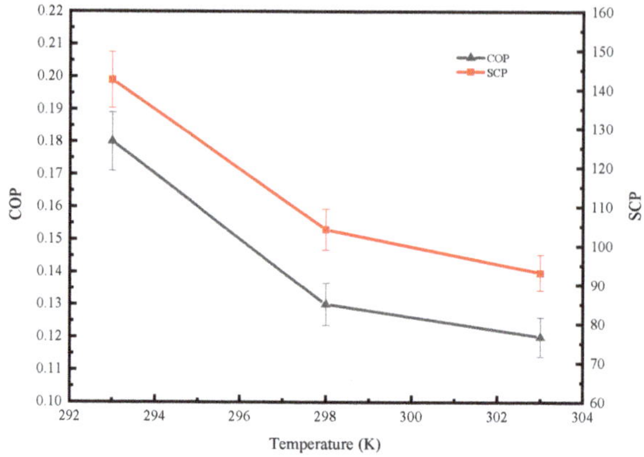

Figure 15. COP and SCP of the system at different adsorption temperatures.

2.7. *Comparison of Experimental Results with Simulation Results*

The experimental results and simulation results are summarized in Tables 2 and 3. The comparison shows that the simulation results basically reflect the variation of system performance under different working conditions. However, as far as the specific values are concerned, the experimental results are slightly lower than the simulation results, mainly because of the following three reasons: (1) The power of the electric heating furnace of the adsorption bed was too large, and the heat generated during the desorption process was excessive, resulting in the system sensible heat increasing and the system performance decreasing. (2) The specific heat capacity selected during the simulation process was mostly fixed, and the specific heat capacity changed with the experimental temperature during the experiment; (3) The leakage heat of the whole experimental system was better. There was more, and the simulation process ignored the heat that the system loses into the environment.

Table 2. Experimental results and simulation results at different evaporation temperatures.

Evaporating Temperature (K)		Q_{ref} (kJ)	COP
288	Simulation result	79.3	0.15
	Experimental result	73.95	0.103
293	Simulation result	121	0.192
	Experimental result	110.9	0.154
298	Simulation result	125.2	0.23
	Experimental result	123	0.172

Table 3. Experimental results and simulation results at different adsorption temperatures.

Adsorption Temperature (K)		Q_{ref} (kJ)	COP
293	Simulation result	142	0.21
	Experimental result	128.2	0.18
298	Simulation result	119	0.19
	Experimental result	93.67	0.13
303	Simulation result	100	0.17
	Experimental result	83.81	0.12

3. Conclusions

1. MIL-101(Cr) was successfully synthesized and the structure of MIL-101(Cr) remained unchanged after $CaCl_2$ was added.
2. With the increase in $CaCl_2$ content, the total pore volume and pore diameter distribution of MIL-101(Cr)/$CaCl_2$ composite material gradually decreased. When $CaCl_2$ concentration reached 30%, the pore volume of MIL-101(Cr) was almost completely blocked by molecules.
3. The COP simulation model showed that the optimal desorption temperature and COP were 351 K and 0.24, 0.1 higher than MIL-101(Cr).

4. Materials and Methods

4.1. Reagents and Materials

Chromium nitrate nonahydrate ($Cr(NO_3)_3{\cdot}9H_2O$, ≥99% purity, Aladdin Biotechnology Co. Ltd., Shanghai, China), terephthalic acid (H_2BDC, ≥99.0% purity), hydrofluoric acid (HF, ≥40% purity, Maclean Biotechnology Co. Ltd., Shanghai, China), *N,N*-Dimethyl formamide (DMF, ≥99.9% purity), ammonium fluoride (NH_4F, ≥99.99% purity), ethanol (≥99.7% purity, Beijing Tongguang Fine Chemical Co. Ltd., Beijing, China), anhydrous calcium chloride (≥99.99% purity), and distilled water (prepared by the laboratory) were used in the experiments.

4.2. Synthesis of MIL-101(Cr)

The parent MIL-101(Cr) was synthesized by hydrothermal synthesis using the procedure reported by Yang [19,20].

4.3. Synthesis of MIL-101(Cr)/$CaCl_2$ Composites

MIL-101(Cr)/$CaCl_2$ composites were prepared by the immersion method [17]. About 0.4 g of MIL-101(Cr) was weighed and dried in an oven at 100 °C for 8 h. The MIL-101(Cr) sample was cooled to room temperature and then immersed in 10%, 20%, and 30% $CaCl_2$ solutions (soaked at room temperature for 2 h). Finally, the soaked material was subjected to centrifugation, washed with deionized water, and centrifuged again to remove excess calcium chloride. The sample was collected and completely dried overnight in a vacuum oven at 373 K.

4.4. Materials Characterization

X-ray diffraction (XRD) was used to analyze the image of the composite. The diffraction peak was used to determine whether the composite was successfully synthesized. The thermogravimetric analysis (TG) was used to test the relationship between material quality and temperature, and its composition and thermal stability were studied. The specific surface area and pore structure of the composite were analyzed by N_2 adsorption–desorption isotherm. The water vapor adsorption isotherm, adsorption kinetics and differential adsorption heat curves of the composite adsorbent were obtained by the gravimetric method, and the huge space and potential in the adsorption refrigeration application were preliminarily determined.

4.5. Measurement of Water Vapor Adsorption Isotherms

The water vapor adsorption isotherm, adsorption kinetics curve and adsorption heat curve of the material were tested by Conta's DVS Advantage water vapor adsorption instrument. A weight about 40 mg of the sample was vacuum dried at 373 K to remove the solvent molecules from the material and adsorb the water vapor into the pores. The material was then cooled to room temperature and placed in the crucible of the instrument, then the interface was clicked on to run the program. During the test, the temperature of the adsorption bed and the crucible were set to 298 K and 308 K, respectively.

4.6. Experimental Rig of Single-Bed Adsorption System with New Type Adsorbent Filling Method

The experimental platform of adsorption refrigeration system adopts the same structure as the previous study [21], as shown in Figure 16. We changed the adsorption bed filling method. Figure 17 shows the adsorption bed fabrication process. We took a 30 cm long stainless steel tube with a diameter of 22 mm, opened a certain number of 5 mm holes, this was the mass transfer hole. The empty stainless steel tube was wrapped with two 300-mesh screens, and clamped to the stainless steel tube with a clamp to prevent the adsorbent powder from being drawn into the pipeline of the system during vacuuming, which affects the normal operation of the experiment. We welded the stainless steel tube with the mass transfer holes into the center of a thick stainless steel tube (adsorption bed) with a length of 30 cm and a diameter of 52 mm. At this time, the inner diameter of the small stainless steel tube was the mass transfer channel of the whole adsorption bed. The composite adsorbent material was then filled into the gap between the thick stainless steel tubes, and the adsorbent bed was welded and sealed and attached to the system via a vacuum ball valve. The adsorption bed had sufficient mass transfer space to enhance the heat and mass transfer capacity of the adsorption bed. The structure of the adsorption bed is shown in Figure 18.

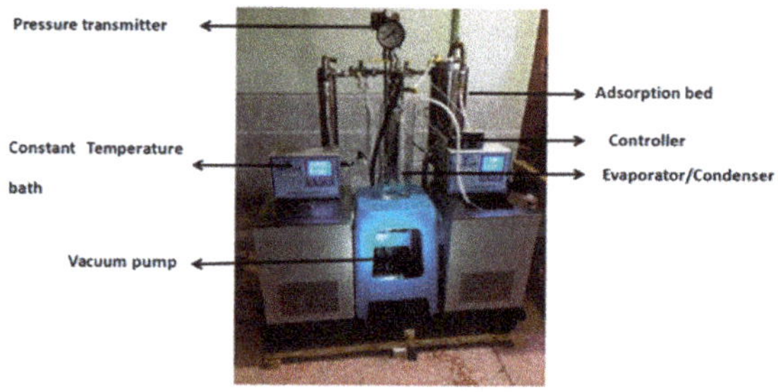

Figure 16. Adsorption refrigeration system performance test device.

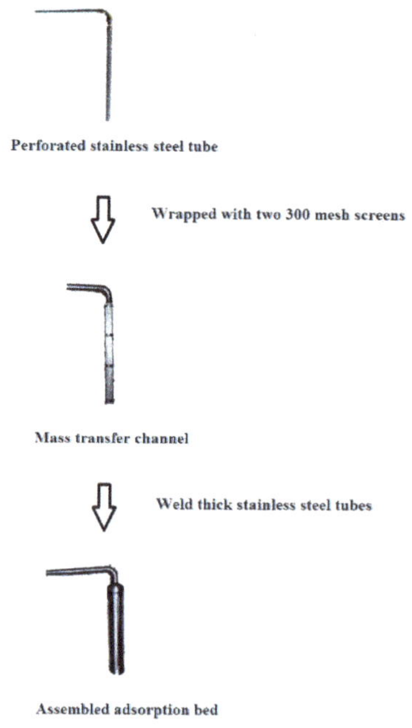

Figure 17. Production process of the adsorption bed.

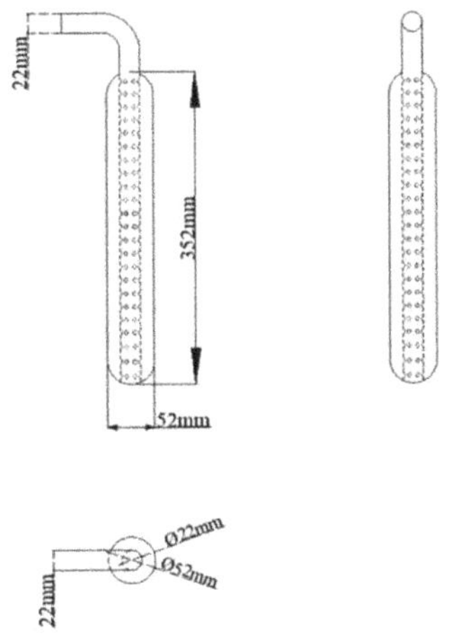

Figure 18. Adsorption bed structure.

4.7. Experimental Procedure for Performance Test of Adsorption Refrigeration System

The schematic diagram of the experimental system is shown in Figure 19. The experimental steps are as follows:

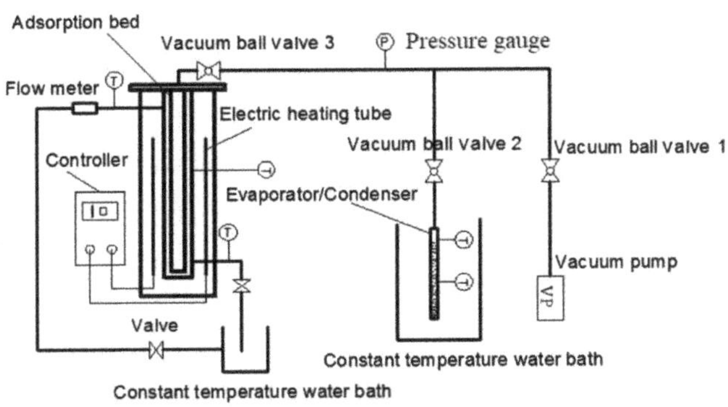

Figure 19. Schematic diagram of adsorption performance test.

Adsorption bed heating vacuum process: because the system requires high vacuum, it is necessary to heat the adsorption bed before the cycle starts, set the heating temperature to 100 °C, then turn on the vacuum pump for 30 s. Then the vacuum ball valve 1 and the vacuum ball valve 3 are opened, and the adsorption bed is heated and evacuated until the internal pressure of the adsorption bed is maintained at about −101kPa. The vacuum ball valve 1 is closed, the constant temperature water bath is turned on, and the adsorption bed is cooled to the adsorption temperature.

To remove excess air from the evaporator/condenser: open the vacuum ball valve 2 on the basis of the previous step and remove excess air from the evaporator/condenser until the pressure in the evaporator/condenser reaches the evaporation temperature. Saturate the pressure, then turn off the vacuum ball valve 1 and then turn off the vacuum pump.

Adsorption–evaporation process: the vacuum ball valve 2 and the vacuum ball valve 3 are opened, and the adsorption bed is connected with the evaporator. The refrigerant in the evaporator changes from a liquid state to a gaseous state and enters the adsorption bed to be adsorbed by the adsorbent, and then the temperature and pressure in the adsorption bed will increase. When the valve is not opened, the pressure in the evaporator remains at the condensation pressure, and the pressure difference between the adsorber and the evaporator is the difference between the condensation pressure and the evaporation pressure. At the beginning of the adsorption process, the pressure inside the evaporator will gradually decrease to the evaporation pressure, and the pressure difference between the adsorber and the evaporator will gradually decrease to zero. At the same time, the water bath that opens the adsorbent bed cools the adsorbent bed to ensure that the adsorbent bed temperature is maintained at the adsorption temperature. The liquid water in the evaporator evaporates into a heat exchange between the water vapor and the external environment, so that the temperature of the evaporator water bath is continuously reduced as the adsorption process progresses, thereby generating a refrigeration effect. The adsorption process ends when the adsorbent bed pressure is the same as the evaporator pressure. The initial adsorption time, changes in evaporator/condenser level, and adsorption end time are recorded.

Heating process: the adsorption bed is heated before the desorption process. The adsorption bed is heated to 100 °C by an electric heating tube, during which the internal temperature and pressure of the adsorption bed are both increased. When the pressure in the adsorption bed is greater than the pressure in the condenser, the temperature rise is completed.

Desorption–condensation process: set the temperature of the water bath where the condenser is set to the condensation temperature and open the vacuum ball valve 2. Due to the pressure difference between the adsorption bed and the condenser, the gaseous refrigerant will enter the condensation through the pipeline under the action of pressure and with the desorption, the pressure difference between the adsorber and the condenser decreases. The condenser condenses into a liquid refrigerant, and the temperature of the condenser water bath increases. When the internal pressure of the condenser is equal to the internal pressure of the adsorbent bed, the condensation process ends.

Cooling process: when the desorption process is finished, the adsorption bed is still in a high temperature and high pressure state, which is not conducive to the next cycle, so the adsorption bed needs to be cooled before the start of the next cycle. The circulating water is opened to the adsorption bed. When the pressure in the adsorption bed drops to the saturation pressure corresponding to the evaporation temperature, the cooling process ends and the next adsorption cycle begins.

The parameters of electric heating tube, pressure transmitter and constant temperature water bath used in the above experiment are shown in Tables 4–6.

Table 4. The main technical parameters of the electric heating tube.

Content	Value
Temperature Range (K)	273–383
Accuracy (K)	0.1
Power Voltage (Hz/V)	50/220
Power (W)	200

Table 5. The main technical parameters of the pressure transmitter.

Content	Value
Power Supply (V)	24
Operating Temperature Range (°C)	−30–80
Signal (mA)	4–20
Range (kPa)	−100–1600
Accuracy (kPa)	0.1

Table 6. The main technical parameters of the constant temperature bath.

Content	Value
Temperature Range (K)	268–373
Temperature Fluctuation (K)	±0.05
Accuracy (K)	0.1
Volume (L)	7.5
Circulating pump flow (L/min)	6
Power Voltage (Hz/V)	50/220

5. Mathematical Model of the Single-Bed Adsorption System

5.1. Basic Assumptions of the Theoretical Model

For an adsorption refrigeration system, its internal energy flow is a complex process, and it is impossible to fully express such a complicated energy flow process by mathematical expression. There is no need to consider such a complicated problem in the engineering application, so the complex problem can be simplified in mathematical modeling, regardless of the heat transfer process of internal load during energy flow, and the lumped parameter method is used to describe the energy flow process. The system parameters and constants involved in the simulation process are shown in Table 7.

Table 7. Physical property parameters used in the simulation.

Symbol	Term	Value	Unit
M_a	Mass of the adsorbent	0.1	kg
M_m	Mass of the adsorbent bed	2	kg
C_{lc}	Specific heat capacity of the refrigerant	4.18	kJ/kg·K
C_m	Specific heat of the stainless steel	0.5	kJ/kg·K
L	latent heat of vaporization of water	2465	kJ/kg

In order to simplify the calculation, the following assumptions were made:

(1) The temperature and vapor pressure inside the entire adsorption bed are uniform.
(2) The water is uniformly adsorbed by the MIL-101(Cr)/$CaCl_2$ composite material, and the water is liquid in the MIL-101(Cr)/$CaCl_2$ composite material.
(3) The pressure difference between the adsorption bed and the condenser are ignored, the pressure difference between the adsorption bed and the evaporator are ignored.
(4) The heat conduction of the casing connected between the adsorption bed and the condenser or the evaporator is ignored, assuming complete insulation between the adsorption bed, the condenser and the evaporator.
(5) In addition to accidental heat exchange between hot water, cooling water and chilled water, the system ignores the cold/heat in the environment.
(6) The specific heat capacity of the MIL-101(Cr)/$CaCl_2$ composite, the specific heat capacity of the adsorbent bed and the specific heat capacity of the refrigerant are constant.

5.2. Mathematical Model of the Basic Cycle

The simulation model of this cycle is based on COP Equation (1).

$$\mathrm{COP} = \frac{Q_{ref} - Q_{eva}}{Q_h + Q_g} \tag{1}$$

The seven heat calculation formulas involved are as follows:

5.2.1. The Sensible Heat Absorbed by the Adsorption Bed During Constant Volume Boost (Q_h)

$$Q_h = \int_{T_{a2}}^{T_{g1}} C_a M_a dT + \int_{T_{a2}}^{T_{g1}} C_{lc} M_a x_{a2} dT + \int_{T_{a2}}^{T_{g1}} C_m M_m dT \quad (2)$$

where T_{g1} is the initial analytical temperature (K), T_{a2} is the adsorption temperature (K), C_a is the specific heat capacity of the adsorbent (kJ·kg^{-1}·K^{-1}), C_{lc} is the specific heat capacity of the refrigerant (kJ·kg^{-1}·K^{-1}), C_m is the specific heat of the metal in the adsorbent bed (kJ·kg^{-1}·K^{-1}), M_a is the mass of the adsorbent (kg), M_m is the mass of the adsorbent bed (kg), and X_{a2} is the adsorption amount of the adsorption bed at the end of adsorption of the adsorption line corresponding to the T_{a2} temperature point. In Equation (2), the first term represents the sensible heat of the adsorbent, the second term represents the sensible heat of the liquid refrigerant, and the third term represents the sensible heat of the adsorbent bed.

5.2.2. Heat Absorbed during Desorption (Q_g)

$$Q_g = \int_{T_{g1}}^{T_{g2}} C_a M_a dT + \int_{T_{g1}}^{T_{g2}} C_{lc} M_a x dT + \int_{T_{g1}}^{T_{g2}} C_m M_m dT - \int_{T_{g1}}^{T_{g2}} M_a h_d dT \quad (3)$$

where T_{g2} is the desorption temperature (K), x is the adsorbent adsorption amount (kg·kg^{-1}), and h_d is the desorption heat (kJ·kg^{-1}). In Equation (3), the first term represents the sensible heat of the cooled adsorbent, the second term represents the sensible heat of the refrigerant, and finally the sensible heat required to cool the adsorbent bed.

5.2.3. The Sensible Heat Taken Away by the Cooled Adsorption Bed (Q_c)

$$Q_c = \int_{T_{a1}}^{T_{g2}} C_a M_a dT + \int_{T_{a1}}^{T_{g2}} C_{lc} M_a x_{d2} dT + \int_{T_{a1}}^{T_{g2}} C_m M_m dT \quad (4)$$

where T_{a1} is the initial adsorption temperature (K), and x_{d2} is the adsorption amount (kg·kg^{-1}) of the adsorbent at the end of desorption corresponding to the T_{g2} temperature point. The first term in Equation (4) represents the sensible heat of the adsorbent, the second term represents the sensible heat of the refrigerant remaining in the adsorbent bed, and the third term represents the sensible heat of the adsorbent bed.

5.2.4. Heat Taken Away by the Cold Source during Adsorption (Q_{ad})

$$\begin{aligned} Q_{ad} = \int_{T_{a2}}^{T_{a1}} C_a M_a dT \ &+ \int_{T_{a2}}^{T_{a1}} C_{lc} M_a x dT \\ &+ \int_{T_{a2}}^{T_{a1}} C_m M_m dT - \int_0^{T_a - T_e} C_{pc} M_a \Delta x dT + \int_{T_{a2}}^{T_{a1}} M_a h_a dT \end{aligned} \quad (5)$$

where C_{pc} is the constant pressure specific heat of the gaseous working fluid (kJ·kg^{-1}·K^{-1}), and ha is the heat of adsorption (kJ·kg^{-1}). The last term in Equation (5) is the sensible heat of the vaporized working gas temperature rising from the evaporation temperature to T_a.

5.2.5. Refrigerating Capacity (Q_{ref})

$$Q_{ref} = M_a L \Delta x \quad (6)$$

where L is the latent heat of vaporization of water(kJ·kg^{-1}) is cyclic adsorption (kg·kg^{-1}).

5.2.6. Heat Released by Condensation (Q_{cond})

$$Q_{cond} = M_a L \Delta x + \int_{T_c}^{T_{g2}} C_{pc} M_a \frac{d_x}{d_t} dT \quad (7)$$

where T_c is the condensation temperature (K). In Equation (7), the first term is the latent heat of saturated vaporization, and the second term is the sensible heat released by refrigerant vapor during condensation.

5.2.7. Sensible Heat Released by Liquid Refrigerant from T_c to Evaporation Temperature T_e (Q_{eva})

$$Q_{eva} = \int_{T_c}^{T_e} C_{lc} M_a \Delta x dT \quad (8)$$

where T_e is the evaporation temperature (K). In fact, as adsorption refrigeration is a complicated heat and mass transfer process, there are various heat losses, so it is very difficult to accurately calculate all kinds of heat. However, it can be theoretically useful to analyze the cycle through these equations.

5.2.8. Expressions for T_{g1} and T_{a1}

Before the beginning of the adsorption process, after opening the valve connecting the adsorber and the evaporator, the gas in the evaporator releases heat from the condensation temperature T_c to the evaporation temperature T_e, and the pressure drops from the condensation pressure P_c to the evaporation pressure P_e. The adsorption amount remains unchanged during the process. According to the ideal gas state equation, its relationship is shown in Equation (9). Before the desorption process begins, there is a preheating process in the adsorbent. The gas temperature in the adsorbent rises from T_{a2} to the initial desorption temperature T_{g1}, while the pressure rises from the evaporation pressure P_e to the condensing pressure P_c, and the gas quantity remains unchanged. The equation is also shown in Equation (10). Similarly, the complete desorption temperature T_{g2} and the initial adsorption temperature T_{a1} satisfy the Equation (11). Equations (12) and (13) are obtained from simultaneous Equations (9)–(11).

$$\frac{P_c}{P_e} = \frac{T_c}{T_e} \quad (9)$$

$$\frac{P_c}{P_e} = \frac{T_{g1}}{T_{a2}} \quad (10)$$

$$\frac{P_c}{P_e} = \frac{T_{g2}}{T_{a1}} \quad (11)$$

$$T_{g1} = \frac{T_c T_{a2}}{T_e} \quad (12)$$

$$T_{g2} = \frac{T_c T_{a1}}{T_e} \quad (13)$$

Author Contributions: Conceptualization, Z.L.; Data curation, B.Z.; Formal analysis, B.Z.; Funding acquisition, Z.L.; Investigation, B.Z.; Resources, Z.L.; Software, F.L.; Supervision, X.Q.; Validation, Y.H., X.Q.; Visualization, Y.H., F.L.; Writing—original draft, B.Z.; Writing—review & editing, B.Z. All authors have read and agreed to the published version of the manuscript.

Funding: This work is supported by the Beijing Natural Science Foundation (Grant No. 3202008), the National Natural Science Foundation of China (Grant No. 51776006) and Dezhou "Blue Fire" project of science and technology development center of Ministry of Education (Grant No. Q9005014201901).

Acknowledgments: Thanks to the laboratory of Environmental Energy college of Beijing university of technology for providing us with the necessary instruments.

Conflicts of Interest: The authors declare no conflict of interest.

Nomenclature

T_{g1}	initial desorption temperature, K
T_{g2}	desorption temperature, K
C_a (T)	specific heat capacity of the adsorbent, kJ·kg^{-1}·K^{-1}
M_a	adsorbent mass, kg
T_{a1}	initial adsorption temperature, K
T_{a2}	adsorption temperature, K
C_{lc} (T)	specific heat capacity of water, kJ·kg^{-1}·K^{-1}
C_m (T)	specific heat capacity of stainless steel, kJ·kg^{-1}·K^{-1}
M_m	adsorption bed mass, kg
Q_h	sensible heat absorbed by the adsorption bed during constant volume boost, kJ·kg^{-1}
Q_g	heat absorbed during the desorption process, kJ·kg^{-1}
h_d	desorption heat, kJ·kg^{-1}·K^{-1}
h_a	heat of adsorption kJ·kg^{-1}
Q_{ref}	cooling capacity, kJ·kg^{-1}
L	latent heat of vaporization of water, kJ·kg^{-1}
Δx	circulating adsorption capacity, kg·kg^{-1}
X_{a2}	water adsorption capacity in gas–solid-phase equilibrium at the time of adsorption in adsorption temperature, kg·kg^{-1}
X_{d2}	water adsorption capacity in gas–solid-phase equilibrium at the time of desorption in desorption temperature, kg·kg^{-1}
Q_{eva}	sensible heat of liquid refrigerant from Tc to evaporation temperature Te, kJ·kg^{-1}
T_c	condensation temperature, K
T_e	evaporation temperature, K

References

1. Wang, D.; Li, Y.; Li, D.; Xia, Y.; Zhang, J. A review on adsorption refrigeration technology and adsorption deterioration in physical adsorption systems. *Renew. Sustain. Energy Rev.* **2010**, *14*, 344–353. [CrossRef]
2. Biplab, C.; Bidyut, B.S.; Pradip, K.C.; Jyoti, P.S. An overview of developments in adsorption refrigeration systems towards a sustainable way of cooling. *Appl. Energy* **2013**, *104*, 554–567.
3. Astina, I.M.; Zidni, M.I.; Hasugian, H.R.; Darmanto, P.S. Experiment of adsorption refrigeration system with working pairs of difluoromethane and activated carbon modified by sulfuric and nitric acids. In *AIP Conference Proceedings*; AIP Publishing LLC: Melville, NY, USA, 2018; p. 020015.
4. Wang, R.Z.; Wang, L.W.; Wu, J.Y. *Theory and Its Application of Adsorption Refrigeration*; China Science Publishing House: Beijing, China, 2007.
5. Zhao, T.; Jeremias, F.; Boldog, I.; Nguyen, B.; Henninger, S.K.; Janiak, C. High-yield, fluoride-free and large-scale synthesis of MIL-101(Cr). *Dalton Trans.* **2015**, *44*, 16791–16801. [CrossRef] [PubMed]
6. Xia, X.; Cao, M.; Liu, Z.; Li, W.; Li, S. Elucidation of adsorption cooling characteristics of Zr-MOFs: Effects of structure property and working fluids. *Chem. Eng. Sci.* **2019**, *204*, 48–58. [CrossRef]
7. Ehrenmann, J.; Henninger, S.K.; Janiak, C. Water Adsorption Characteristics of MIL-101 for Heat-Transformation Applications of MOFs. *Eur. J. Inorg. Chem.* **2011**, *2011*, 471–474. [CrossRef]
8. Rezk, A.; Al-Dadah, R.; Mahmoud, S.; Elsayed, A. Experimental investigation of metal organic frameworks characteristics for water adsorption chillers. *Proc. Inst. Mech. Eng. Part C J. Mech. Eng. Sci.* **2012**, *227*, 992–1005. [CrossRef]
9. Rezk, A.; Al-Dadah, R.; Mahmoud, S.; Elsayed, A. Investigation of Ethanol/metal organic frameworks for low temperature adsorption cooling applications. *Appl. Energy* **2013**, *112*, 1025–1031. [CrossRef]
10. Shi, B.; Al-Dadah, R.; Mahmoud, S.; Elsayed, A.; Elsayed, E. CPO-27(Ni) metal–organic framework based adsorption system for automotive air conditioning. *Appl. Therm. Eng.* **2016**, *106*, 325–333. [CrossRef]
11. Teo, H.W.B.; Chakraborty, A.; Kayal, S. Post synthetic modification of MIL-101(Cr) for S-shaped isotherms and fast kinetics with water adsorption. *Appl. Therm. Eng.* **2017**, *120*, 453–462. [CrossRef]
12. Garzon-Tovar, L.; Pérez-Carvajal, J.; Imaz, I.; Maspoch, D. Composite Salt in Porous Metal-Organic Frameworks for Adsorption Heat Transformation. *Adv. Funct. Mater.* **2017**, *27*, 1606424. [CrossRef]

13. Yan, J.; Yu, Y.; Ma, C.; Xiao, J.; Xia, Q.; Li, Y.; Li, Z. Adsorption isotherms and kinetics of water vapor on novel adsorbents MIL-101(Cr)@GO with super-high capacity. *Appl. Therm. Eng.* **2015**, *84*, 118–125. [CrossRef]
14. Aristov, Y.; Restuccia, G.; Cacciola, G.; Parmon, V. A family of new working materials for solid sorption air conditioning systems. *Appl. Therm. Eng.* **2002**, *22*, 191–204. [CrossRef]
15. Wang, J.Y.; Wang, R.Z.; Wang, L.W. Water vapor sorption performance of ACF-$CaCl_2$ and silica gel-$CaCl_2$ composite adsorbents. *Appl. Therm. Eng.* **2016**, *100*, 893–901. [CrossRef]
16. Chan, K.C.; Chao, C.Y.; Sze-To, G.; Hui, K.S.; Chao, C.Y.H. Performance predictions for a new zeolite 13X/$CaCl_2$ composite adsorbent for adsorption cooling systems. *Int. J. Heat Mass Transf.* **2012**, *55*, 3214–3224. [CrossRef]
17. Permyakova, A.; Wang, S.; Courbon, E.; Nouar, F.; Heymans, N.; D'Ans, P.; Barrier, N.; Billemont, P.; De Weireld, G.; Steunou, N.; et al. Design of salt–metal organic framework composites for seasonal heat storage applications. *J. Mater. Chem. A* **2017**, *5*, 12889–12898. [CrossRef]
18. Liu, Z.; Chen, Y.; Sun, J.; Lang, H.; Gao, W.; Chi, Y. Amine grafting on coordinatively unsaturated metal centers of MIL-101Cr for improved water absorption characteristics. *Inorg. Chim. Acta* **2018**, *473*, 29–36. [CrossRef]
19. Elsayed, E.; Wang, H.Y.; Anderson, P.A.; Al-Dadah, R.; Mahmoud, S.; Navarro, H.; Ding, Y.L.; Bowen, J. Development of MIL-101(Cr)/GrO composites for adsorption heat pump applications. *Microporous Mesoporous Mater.* **2017**, *244*, 180–191. [CrossRef]
20. Yang, J.; Zhaoa, Q.; Li, J.; Dong, J. Synthesis of metal–organic framework MIL-101 in TMAOH-Cr(NO3)3-H2BDC-H2O and its hydrogen-storage behavior. *Microporous Mesoporous Mater.* **2010**, *130*, 174–179. [CrossRef]
21. Liu, Z.; Zhao, B.; Zhu, L.; Lou, F.; Yan, J. Performance of MIL-101(Cr)/Water Working Pair Adsorption Refrigeration System Based on a New Type of Adsorbent Filling Method. *Materials* **2020**, *13*, 195. [CrossRef] [PubMed]

Sample Availability: The MIL-(Cr)/$CaCl_2$ composites covered in this article can be obtained from us.

MDPI AG
Grosspeteranlage 5
4052 Basel
Switzerland
Tel.: +41 61 683 77 34

Molecules Editorial Office
E-mail: molecules@mdpi.com
www.mdpi.com/journal/molecules

www.ingramcontent.com/pod-product-compliance
Lightning Source LLC
Chambersburg PA
CBHW062040220926
51850CB00010B/2617
9783725868940